TECHNOLOGIES
WITHOUT
BOUNDARIES

TECHNOLOGIES WITHOUT BOUNDARIES

On Telecommunications in a Global Age

Ithiel de Sola Pool

Edited by
Eli M. Noam

Harvard University Press
Cambridge, Massachusetts
London, England 1990

This book is printed on acid-free paper, and its binding materials
have been chosen for strength and durability.

Library of Congress Cataloging-in-Publication Data

Pool, Ithiel de Sola, 1917–1984
    Technologies without boundaries : on telecommunications in a
global age / Ithiel de Sola Pool ; edited by Eli Noam.
        p.   cm.
    Includes index.
    ISBN 0-674-87263-0
    1. Telecommunication—Social aspects.   2. Telecommunication—
Technological innovations.   I. Noam, Eli M.   II. Title.
HE7631.P66   1990   90-37817
302.2—dc20   CIP

Preface

Ithiel de Sola Pool died in March of 1984. Among the papers he left behind was a massive manuscript which became this book, *Technologies without Boundaries*. In his last years, Pool had worked hard to summarize his insights about communications and society. The most notable result was *Technologies of Freedom* (1983), a book which became influential, far beyond the circle of communications scholars, for its insight, clarity, and broad perspective. For this achievement Pool was the posthumous recipient of the 1984 Gladys M. Kammerer Award of the American Political Science Association for the best book in the field of U.S. national policy. Only a handful of people knew that these six chapters focusing on the legal status of American media were only one third of a larger whole. The rest was intended as the second volume; but no time remained to finish the task.

In 1988 Harvard University Press asked me to edit the surviving draft. My feelings were mixed. I felt honored to be entrusted with a magnum opus by one of the major scholars of communications. But I had not been one of Pool's students and colleagues and had actually met him only twice, since my own entry into the field came during the period of his illness. In contrast, there were many accomplished Pool students and disciples with a far better claim to interpret his views. Yet to conclude this book was also a personal and professional duty. So I took on this pleasing obligation.

To appreciate Pool's contribution, one should know something about the man. Ithiel de Sola Pool was an outstanding political scientist and a pioneer in the field of social and political communication. Born in 1917 in New York as the son of the distinguished rabbi and scholar of the Spanish and Portuguese Synagogue, Pool studied at the University of Chicago during the Hutchins era. He participated in the radical student politics of the time and acquired a life-long and unwavering support for democracy and opposition to totalitarianism of all stripes. After conducting major research at Stanford on political symbols, in 1953 he joined the Massachusetts Institute of Technology, where he served as director of the Research Program on Communications Policy and as chairman of the Political Science Department. As a scholar, he became a leading contributor to an understanding of political elites, of communications processes in society, and of the transmission of political values across time and space. Pool was honored by the American Political Science Association in 1964 when his *American Business and Public Policy* (with Raymond Bauer and Lewis Dexter) received the Woodrow Wilson Foundation Award for the best book of the year published on government politics or international affairs. The book advanced understanding of political behavior of congressional and pressure groups by analyzing the communications flows in the political process. Similar classics are his *Handbook of Communications* (1973, with W. Schramm et al.) and the edited volume *The Social Impact of the Telephone* (1976).

Pool was also a methodological innovator, one of the first to apply computer-based quantitative methods to classical political theory. He developed a computer simulation model of voting behavior and used it during the presidential campaign of John F. Kennedy. Pool was an advisor to successive administrations and to many international organizations. His work on development processes and modernization brought him into close ties with many scholars and academic institutions in the Third World. Pool had a deep interest in understanding the problems of revolutions, including the one in Vietnam, and in the social conditions, particularly the mass communications, likely to influence their success. He was ahead of his time in anticipating how technology would give individuals an opportunity to participate in the political process.

Technologies without Boundaries extends Pool's analysis into the international realm. He presents his vision of a new world resulting from the social, political, and cultural consequences of communications technology. His perspective on technology is that of the social scientist, rooted in a broad understanding of history and political forces.

Pool shows how new communications technologies were often controlled by governments at first, but that control eventually gave way to freedom and diversity. He discerns five trends whose confluence greatly affect tomorrow's communications: distance ceases to be a barrier to communication; different modes of communications merge with one another; communications and computing overlap; information activity becomes a major undertaking of advanced economies; and mass media become individualized. While all of these trends have been noted before, Pool's contribution is his analysis of their ramifications. Though disavowing an attempt at prognostication, he speculates about the long-term impact of these trends on societies and nations. He concentrates on four main areas: changes in the spatial patterns of economic and social activity; threats to liberties; new forms of social cohesion and individuality; and environmental implications.

Pool has described himself a "soft" technological determinist— one who sees technology as enabling and as creating powerful currents but not necessarily controlling them. Policy, economics, culture, and the social context affect technology, too. Causality is mutual. The same technology may cause opposite effects. Whether information media enhance or retard freedom and culture depends on the rules under which they are allowed to operate.

Pool does not distrust technology but governments. In his view, not computers but policy threatens freedom. His concern is that governments, fearful of a loss of control over sovereignty and culture, will continue to resist opening new communications channels, despite centuries of experience with the self-defeating results of such restrictiveness. His theme is liberty. He is less concerned with private power, the subject of many other authors. Is this a blind spot? Pool's earlier work shows great awareness of the potential for corporate abuse. Yet he does not dwell on it here. In communications, he viewed the cure of control as worse than the malady. Yet market failure is not a hypothetical construct, and it can legitimize at least some protective policies even on the grounds of freedom of speech.

For example, Pool sees media policy as generally reducing diversity in favor of centralism. But the elimination of the legal barriers which he decries may not always be sufficient to establish a system of many voices. A governmental role in *adding* to program supply seems therefore legitimate—for example, through support of nonexclusive public broadcasting institutions, or of programs that would otherwise not be produced.

Thus Pool may be at times overly negative about any role for government policy in achieving diversity. For example, in telecommunications, where the American environment now has become a loose federation of multiple and often competing carriers, some physical interconnection and content-access arrangements are often necessary to assure the flows of communication. Thus, when the New York State Public Service Commission recently established a system of content-neutral "common carriage rules" for telephone carriers, open to all program providers without license, censorship, and control, we were directly inspired by Pool's concepts, but not by his view that government serves primarily to restrict the rivals of established media.

Editing a posthumous manuscript is never easy. One becomes an executor—and sometimes executioner—of another person's thoughts, without the freedom of one's own creation. Occasionally I disagreed with Pool's views, but I did not endeavor to censor them. However, in instances where the facts have changed, I assumed he would have consented to correction. The original manuscript was edited to reduce the inevitable repetitions that creep into an early draft and to eliminate unneeded specifics where development has moved on. Since this book is not a reference text, I chose not to update the book's sources or data into the year 1990 but instead to concentrate on uncluttering the essential Pool.

Technologies without Boundaries is likely to cast a wide influence. It is a specialist's book for generalists, and a generalist's book for specialists. Its scope is the world, culture, and the future, and its themes are as fresh and relevant today as when Pool wrote about them. If anything, the collapse in the past decade of dictatorships around the world has made this book even more contemporary than it was at the time of its writing.

Eli M. Noam

Acknowledgments

Were Ithiel de Sola Pool alive today, he would surely wish to express his heartfelt thanks to the many colleagues and students at the Massachusetts Institute of Technology with whom he worked while writing the manuscript and those with whom he interacted while traveling around the world on his sabbatical. These include particularly scholars at the East West Center and the University of Hawaii; the Institute for Communications Research, Keio University; the Research Institute for Telecommunications and Economics, Tokyo; and Churchill College, Cambridge University. Samuel Popkin helped shepherd the manuscript through the review process. David Park assisted in the early editing; Theresa Bolmarcich and Michael McManus were helpful in the later stages. The John and Mary Markle Foundation generously provided financial support for Pool's research. And finally, at Harvard University Press, Arthur Rosenthal, Michael Aronson, and Susan Wallace accomplished with grace and skill the enormous task of orchestrating this project.

EMN

Contents

PART II   SATELLITES, COMPUTERS, AND GLOBAL RELATIONS

**PART III ECOLOGY, CULTURE, AND
COMMUNICATIONS TECHNOLOGY**

PART I

COMMUNICATIONS AND THE CHANGING ENVIRONMENT

Chapter 1

From Mass Media Revolution to Electronic Revolution

In the last third of the twentieth century we have reached a historical corner. Currents in industrial societies that have flowed one way for centuries are reversing direction. Since the seventeenth century, modernization has meant the growth of factories, cities, freedom of expression, nationalism, and mass culture. But from the communications technologies of the past few decades may sprout, as from windblown seeds, consequences that could not have been anticipated at the dawn of the electronic era. We may begin to see spatial organizations of human settlement quite unlike the classical city; threats to freedom of speech and of the press; erosion of the sovereign state; and a fracturing of society's cohesion.

In this book we will consider how new communications technologies are producing such social changes. But first it would be useful to review an earlier communications revolution—one in which, because it is past, we may better perceive how much technology of communication has shaped our lives.

The Example of the Printing Press

It all started in China, where the making of paper from textiles was developed early in the second century A.D. Paper production spread westward with the Arab conquest of Turkestan in 751,[1] and by the twelfth century paper was manufactured in Spain and by the thirteenth in Italy. Paper's importance lay in the fact that it was much cheaper than the parchment on which medieval manuscripts had been written; around 1400 the cost of paper in Italy was one-sixth that of parchment.

Movable type was also a Chinese invention, attributed to Pi

Sheng and the eleventh century.[2] It did not become very important in Imperial China, however, because that society required only the frequent reprinting of a relatively small number of classics, and for that purpose printing from full page blocks was satisfactory. In Korea, however, movable type was adopted and developed for more general use. In 1241 metal type replaced Pi Sheng's earthenware characters. In 1403 a type foundry existed in Korea, and King Tha-tjong proclaimed: "It is our will and law that type shall be produced from copper and that various books be printed, so that in this way knowledge may be more widely disseminated for the countless needs of all."[3] Gutenberg innovated, as well as combined, these printing techniques. For example, because Korean type consisted of flat squares, it could not be firmly bound together in a frame. Gutenberg put his typeface on the top of a tall cube of metal; this added half inch of base permitted the composed page in its clamps to become a solid block for handling.

Thus, some of the inventions that came together to make what we call printing had been in the process of adoption for up to a millennium. Once the system was introduced in Europe, it spread rapidly. By the 1490s the larger states had at least one publishing center.[4] Between 1481 and 1501, 268 printers in Venice turned out two million volumes.[5]

What impact did this early printing activity have on the society of the day? Paul Lazarsfeld once jokingly suggested that social researchers (if such had been around), evaluating printing a decade or so after its invention, would have concluded that the new device was not very significant. Scribes, he noted, had already been efficiently producing the important books, and the new printers produced mainly the same old texts, such as the Bible, which were already readily available to the tiny literate minority. It is true that hand copying continued to be competitively viable throughout the fifteenth century; in the region of Paris and Orleans alone, there were about 10,000 scribes at work.[6] Often they copied printed books; when a typical edition of 200 to 1,000 of an early printed book ran out, it was more economical to meet residual demand by hand. So, like many innovations, printing did not at first change society very much. Yet none of us now doubts the profound effects that printing had in the end. And, indeed, they were rapid ones. The mere fact that a printer could produce on the average not quite one volume a day, while a scribe produced two a year, made long-term change inevitable.[7]

A brilliant analysis of the social changes that followed printing has been offered by Elizabeth Eisenstein.[8] Here are some of the developments she notes:

Growth of Protestantism. Family Bibles became available to common people; priests were no longer needed as interpreters. Tracts, sermons, and opinions of all sorts, often controversial, were diffused in print. Manuscript copying had been one of the economic mainstays of the monasteries. Printing, on the other hand, was done by bourgeois craftsmen. That displacement of jobs from the domain of the Church to that of the guilds was one force for a shift in the balance of power.

Growth of censorship. In reaction to the heresies that flowed from the printshops, the Church tightened censorship and controls. In 1501 a bull was issued by Pope Alexander VI against the unlicensed printing of books. In 1559 the Index Expurgatorius was begun. Governments, too, reacted to the "menace" of the printed word. In 1556—eighty years after its introduction into England by William Caxton—the British government placed printing under the charter of a Stationer's Company. Twenty-seven years later the Star Chamber restricted the right to print to the two universities and the existing 21 shops in the City of London. It granted the Stationer's Company power to inspect printing offices and to seize and destroy offending documents and presses.[9]

Restriction of domestic industry. Those countries that restricted printing lost publishing activity to those that left it free. British controls on the founding of type in 1637 made that country dependent on the Dutch for such devices. French controls on printing in the sixteenth century, and particularly the burning at the stake of Etiénne Dolet in 1546, caused many printers to flee to Holland. In 1640 Richelieu had to send there for printers to open a French royal printing plant.

Rise of libertarian urges. These restrictions on their activity made printers rebellious. In the years before the revolutionary mob tore down the Bastille in 1789, over 800 authors, printers, and book dealers had been incarcerated there.

Codification of law. Printing changed the practice of law and the way bureaucracies functioned. Before printing, exact texts of statutes or legal decisions were generally to be found only at the central court. Local authorities had to rely on a remembered sense of the law, with the result that an oral or common law, characterized by local variation and consequent autonomy, was the basis for judgments. After printing was introduced, a precise text was available at every local court. Local autonomy declined, and centralized nation-states were fostered.

Development of the concept of intellectual property. Print technology also led to new notions of intellectual property and to the concept of copyright, because at the printing press the number of copies produced could be controlled and counted in a way that was impossible with hand copying.

Development of national cultures. Medieval scholars had been itinerant, wandering from monastery to monastery in search of learned manuscripts. After printing came into use, they became sedentary, and the world of learning split into national domains.

Proliferation of disciplines. The opportunity to publish was an incentive to write; with an audience available, ego drew many to authorship. The new scholars were less likely to study a few great texts and more likely to indulge in specialized writing of new books on their own. Much of this new writing was, of course, inferior to the few classics that had been winnowed out into our permanent heritage. Among the new books, though, were such useful (but unclassic) things as handbooks, lawbooks, astronomical tables, textbooks, bibliographies, memoirs, essays, stories, travelogues, and gossipy social commentaries.

Growth of mysticism. Among the popular publications were many on the arcane. Mysticism was a counter-culture raised from an oral to a written tradition.

Growth of science. Various sciences and specialties, such as medicine, astronomy, engineering, and navigation, were aided in the same way, that is, by having their lore made available in print.

Rise of the idea of history and progress. Historical publishing provided better records of the past. The idea of progress might not have arisen if vagueness about the past allowed people to forget how very different things had once been.

Emergence of modern languages. Publishing of popular and secular materials led to the codification of the vernaculars into the modern European languages.

Separation of the sacred and secular. With the appearance of printed news media, sermons in church no longer were a major means of reporting community news. As church services became more purely religious, the notion of a separation of secular and sacred spheres became clearer.

Growth of literacy. Printed matter was an incentive to learn to read. The literate person could become a reader to his community without being a professional bard or a priest.

The disciplining of children. The need to teach literacy transformed childhood. Many children, and eventually all of them, were forced to spend much of their time in a highly disciplined, achievement-oriented, book-based institution—the school. Childhood became a period of training and aspiration. Vast institutions were built for teaching. Secular schoolteaching became a major profession.

Creation of new professions. Other reading and writing professions, such as journalist, editor, printer, and librarian, evolved as a result of the print revolution. These occupations were held primarily by members of the new middle class.

With the benefit of hindsight into these developments of the sixteenth and seventeenth centuries we can better see the ways in which a change in communication technology affects many aspects of life and society.

The Era of Digital Communication

A revolution in communications technology is taking place today, a revolution as profound as the invention of printing. Communication is becoming electronic. For untold millennia, man, unlike any other animal on earth, could talk. Then, for about 4,000 years, mankind also devised ways to embody speech in a written form that could be kept over time and transported over space.[10] Then, with Gutenberg, the third era began, and for the past five hundred years written texts could be disseminated in multiple copies. In the last stages of this mass media revolution the phonograph, photograph, tape recorder, and movie camera made it possible to copy and distribute voice and pictures.

We are now entering a fourth era ushered in by a revolution of comparable historical significance to that of print and the mass media. We have discovered how to use pulses of electromagnetic energy to embody and convey messages that up to now have been sent by voice, picture, and text. Just as writing made possible the preservation of an intellectual heritage over time and its diffusion over space, and as printing made possible its popularization, this new development is having profound effects on civilization. Five aspects of electronic communication, already visible in the last quarter of the twentieth century, are likely to change society as much as printing did five centuries earlier.

1. Distance is ceasing to be a barrier to communication. As a result, the spatial organization of human activity will profoundly change.
2. Speech, text, and pictures are being represented and sent by the same kind of electrical impulses, a common digital stream. Separation of these modes is diminishing.
3. In this "information society," a greatly expanded proportion of all work as well as leisure is being spent in communication. Information handling is a growing portion of all of human activity.
4. Computing and communication are becoming one, which is to say that communicating and reasoning are being reunited. With messages converted into electronic bits, they may be not only electronically transmitted but also manipulated by logical devices and transformed.
5. The mass media revolution is being reversed; instead of identical messages being disseminated to millions of people, electronic technology permits the adaptation of electronic messages to the specialized or unique needs of individuals.

These five changes and their ramifications are the subject of this book. While some of the effects to date look modest (as those of printing would have looked after its first few decades), their ultimate impact will be profound. The following chapters will offer a description of processes already unfolding and will speculate about extension of those trends. But there are no imagined futures laid out here, only clues to how to think about them.

Early communications technologies

The seeds of the electronic revolution can be traced to late in the eighteenth century, even before the explosion of the mass media began. As soon as eighteenth-century scientists found that electric currents would travel to a distance, they began to speculate about the use of these currents for signaling devices. Several scientists invented a telegraph in one form or another. A clumsy early device was made by Samuel von Sömmering, who in 1809 strung 26 pairs of wires, one for each letter of the alphabet, between two rooms. Whenever a circuit was closed in the transmitting room, electrolysis started in one of 26 water jars in the other room, and bubbles would rise. By watching the sequence of bubbling jars the receiver could read an alphabetic message. This device proved too slow for practical use.

To Samuel Morse goes credit not for the *idea* of a telegraph, which was already well understood, but for launching in 1844 a practical technology based on the opening and closing of an electrical circuit according to a code that bears his name. Various nineteenth-century scientists developed more advanced devices that sent electromagnetic signals decipherable by retransformation. In 1850 F. C. Bakewell developed a facsimile device; the current went on or off as a lever scanned the lines or blanks of a drawing. In 1876 Alexander Graham Bell invented the telephone, in which electric current reproduced the modulations of the voice. Two years later he invented a "photophone," in which light, rather than electricity, was the carrier. In 1890 Heinrich Hertz identified what we now call radio waves; and, just as with electricity earlier, it was immediately obvious that these might be put to practical use for signal transmission. Guglielmo Marconi dedicated himself to devising a way of doing this. In 1896 he patented a device for generating and detecting hertzian waves.

In the twentieth century the growth of electronic communication has been exponential. By 1939 the number of telephone calls in the United States already exceeded the number of letters mailed.[11] Radio broadcasting began in 1921. Between 1950 and 1965 the percentage of American homes that contained a television set rose from 12.3 percent to over 90 percent.

It was also in the 1950s that the computer came into use. For our purpose one must think of the computer not as a stand-alone machine used to do calculations but in its more general sense as a class of devices that perform logical functions. These are increasingly common: in washing machines they control the cycles; in TV sets they improve the tuning; on our wrists they measure time;

under the hoods of cars they control combustion; in offices they turn typewriters into word processors. And many of them are connected up to form parts of communication systems.

Digital communication has many significant technological features, which are explained in Chapter 2. Suffice it to say here that with digital devices all forms of electronic communication become easily transmittable through the same conduit channels and also become readily manipulable by computers. Only with these developments could the revolutionary implications of electronic communication take shape. The telegraph, telephone, radio, and television, each a wonderful device in its own field, can all be seen as extensions of the massive diffusion of communication that began with Gutenberg. With them, the conquest of distance and the growth in information activities were already being achieved. But development of a unified carrier system for all means of communication, the possibility of logical manipulation of messages, and the resulting possibility of individualizing them all require digital information processing, and it is these developments that have profound implications for the future.[12]

Information and the Second Industrial Revolution

The electronic revolution is in one respect a continuation of the mass media revolution. Both are part of that great historical process in which work by hand is replaced with work by brain. The human role in production becomes that of information processor, reaching decisions and giving instructions. Work becomes the moving and processing of signals, not objects.

This fact has been flagged by various writers, who have noted the growth of what the geographer Jean Gottmann has named the quaternary sector.[13] Economists call agriculture and other processes that create the raw materials for human subsistence the primary sector; the transformation of those goods by manufacture is the secondary and the provision of services the tertiary sector.

Until the industrial revolution, the overwhelming majority of all workers were in the primary sector. Farm families produced only a small surplus over what they themselves consumed, and most of the world's people had to farm. The term "industrial revolution" hides the fact that one of its greatest achievements was to increase agricultural productivity. With the rise of food production, as well as with the growth of industries and cities, one advanced country after another reached the point in the twentieth century where more than 50 percent of the working population were in the secondary sector.

But then a reversal set in which Daniel Bell has designated the postindustrial society. The proportion of workers in manufacturing began to decline and the tertiary or service sector started to grow. With assembly lines and automation, it no longer took a growing number of workers to produce the increasing volume of goods that an ever more affluent society consumed. Employment in manufacture fell, while employment in the less well automated service sector grew. A larger and larger proportion of the population were not producing goods but were providing transportation, security, health, and other services.

As the tertiary or service sector rose, researchers noted contradictory processes within it. While the service sector in aggregate kept growing, still in the postindustrial society many services were becoming less and less available. Household servants were disappearing; so were chauffeurs; and laundries, like factories, were becoming automated. What was growing in the 1950s and 1960s were information processing and management activities. A new distinction was therefore introduced. The activities that handled symbols were designated the quaternary or information sector and distinguished from other services.

Numerous scholars have pointed to the remarkable growth of this quaternary sector. Marc Porat found that in 1900 only 10 percent of U.S. labor force was engaged in the information sector; it grew to 27 percent in 1960, 48 percent in 1970, and over 50 percent in the late 70s.[14] This rate of growth has decreased since the 60s, as eventually it had to do. The growth during the 60s was truly extraordinary and can only be described as a silent revolution. In one decade the proportion of the labor force working with paper went from one fourth to one half. That is a revolution of speed and magnitude unprecedented in human history. Ironically, it happened during a time when radical students in Paris, Beijing, and Cambridge, Massachusetts, were proclaiming revolution. All the while, little noticed, a much more fundamental upheaval than the one they had in mind was taking place.

The mass media revolution of the past 150 years can be viewed as an aspect of the mass production that came with industrialization.[15] The craftsman-printer who produced one page at a time by hand, for a total of 2,000 sheets in a ten-hour day, was displaced by the power press, and then the rotary press (adopted by the London *Times* in 1814) and a series of similar inventions. Before then, 5,000 copies had been a good circulation for an American newspaper. The *New York Sun*, one of the first penny newspapers, reached a circulation of 27,000 by 1835 when it was two years old, and *The Herald* reached 40,000 by 1836.

By the mid-twentieth century an issue of a major American metropolitan newspaper might reach as many as a million people nationwide (or in the case of *Pravda* in the Soviet Union, 7 million). A mass magazine, such as *Reader's Digest*, might even sell 18 million copies and reach 40 million readers at low price. Even so, a top American television network broadcast could reach 80 million viewers, at a per capita cost substantially less than that of a newspaper. These low costs were made possible by centrally producing identical messages in great information factories and distributing them to millions of people.

A second industrial revolution has begun to change the emphasis on mass production, both for commodities and messages. Power machinery, mass production, and mechanics were the key to the first industrial revolution, whose hallmark was cost reduction. The keys to the second industrial revolution have been chemistry and then electronics, and its emphasis is diversification. The standardization of the assembly line was relaxed as computer-controlled production allowed efficiency within the diversity which an affluent society demanded.

The trends of the second industrial revolution have been the same in the media field. Computer-controlled composition has allowed newspapers and magazines to appear in local and specialized editions containing different ads and different features. General magazines such as *Life* and *Look* have folded, while specialized magazines are thriving. Cable television allows many more channels than over-the-air television can provide. Cassettes, information retrieval devices, and various other technologies are reinforcing the trend toward diversity, and so is consumer demand. We are, perhaps, at the end of an era. The benefits and penalties of mass production and mass communication—brought by the great industrial revolution—may not be the benefits and penalties of the future. The new electronic revolution in industry and in communication may be spawning a new set of problems along with new advantages.

The Consequences

If the technological trends are indeed as described, the consequences both for good and ill will be seen in many aspects of life. Among those consequences we will focus on but a few, specifically on consequences in four main areas: (1) changes in the spatial patterns of human activities; (2) threats to freedom; (3) changes in the balance between social cohesion and individuality; and (4) implications for environmental conservation.

(1) Changes in spatial patterns of human activities. If distance ceases to place any significant economic burden on communication, then both nations and cities—the frames of most human activity today—will be transformed. Except insofar as governments enforce that constraint, human interaction in a world where its cost is no longer a significant bar need not be confined within national boundaries. Many problems that today are domestic will have to be dealt with internationally, with all the added difficulty in resolving conflicts among nations. Information services, whether retrieval systems, computers, data processing, or entertainment, will become available from outside the national domain at costs hardly greater than from within. As happened four hundred years ago when governments tried to restrict printing, governments that try to stop the international electronic flow of information by mercantilist policies will find that they pay a considerable price in productivity for doing so; they will lose out to competing countries that allow free use of any information.

(2) Effects on the freedom of speech. The cosmopolitan world of free-flowing electronic information that technology is making possible has many attractions, both cultural and economic, but governments alarmed about national sovereignty, protection of their cultures, or even more often about protection of their prerogatives are not likely to view it with favor. Their response is likely to be increased restriction and repression in a futile attempt to limit free international communication.

In the past, governments reacted—unsuccessfully—to new

problems that arose with print by imposing censorship and controls; that may be a clue to the future. The Western tradition of freedom of speech and press developed in an era when there was no electronic communication. New electronic media then posed special problems. Most Western countries adopted a trifurcated system: one regime for press, another for broadcasting, and a third for common carriers.

In the United States, the regime that applies to print is governed by the words of the Constitution that "Congress shall make no law . . . abridging the freedom of speech, or of the press." In the traditional domains of communication those words are taken literally. As Justice Hugo Black expressed it, "no law" means "no law."[16] Special taxes on the press are unconstitutional; so is imposing licensing of it; so is imposing a compulsory right of reply.

Under the regime applied to broadcasting, because of the restrictions on spectrum, the broadcaster is considered a trustee, licensed to serve the public interest. Even where the facilities and decisions are his, he must exercise those decisions in a way that provides a responsible public forum. Licenses, special fees, right of reply, and public review of content, all anathema to freedom of the press in the print tradition, were all accepted even for private broadcasting.

Alongside the print model and the broadcast model of a communication system, a third model evolved through America's two centuries. That was the common-carrier model. Even without monopoly, a carrier should serve all comers without discrimination in order to facilitate communication. The carrier may be restricted in ancillary businesses in which it could become its own favored customer. Most important of all, the carrier should pay no attention to the content of what it carries. It should read no letters, monitor no phone calls.

In most countries both the postal and the telecommunications carriers are government-owned. In the United States they are private, though licensed and regulated by the state.

If the lines separating publishing, broadcasting, cable television, and the telephone network are broken in the coming decades, then which of the three regulatory models will dominate public policy? There is bound to be a great debate, and sharp divisions between conflicting interests. Will regulation to protect the public interest begin to extend over the conduct of the print media as they increasingly use regulated electronic channels? Conversely, will concern for the traditional notion of a free press lead us to free the broadcast media and common carriers from regulation? Will content-neutral common carriage become a thing of the past, or a governing principle of communication conduits and interconnecting networks in the future?

The system of free public discourse is at stake in a debate over policies that are only tangentially related to freedom. The debate may take place in terms of technical efficiency and the protection of economic institutions, but its outcome may determine how much political freedom will exist some decades hence. Technology is spurring the debate; it alone need not shape the outcome. But unless decision makers understand the implications well, then technology and economic interests may do just that.

(3) Social cohesion versus individuality. Besides the prospect of a vast geographic restructuring of activities and of major challenges to the tradition of freedom, the second industrial revolution is likely to have a significant impact on the cohesiveness of modern societies. Only a few years ago a popular criticism of contemporary society was its trend toward conformism under the influence of mass media. During the nineteenth century, beginning with the rise of the penny papers, more and more people came to participate in an increasingly standardized mass culture. Regional and local cultures declined. Entire nations, and even to a degree the world, came to be absorbed in the same few television shows, the same movies, and the same popular magazines; people everywhere read and saw essentially the same world news from the same few news services.

But suppose the trend toward homogenization has been reversed. What will it mean if audiences are increasingly fractionated into small groups with special interests? What will it mean if the agenda of national fads and concerns is no longer effectively set by a few mass media to which everyone is exposed? Such a trend raises for society the reverse problems from those posed by mass conformism. The cohesion and effective functioning of a democratic society depends upon some sort of public agora in which everyone participates and where all deal with a common agenda of problems, however much they may argue over the solutions. If that agora in the media were to disappear, the same social critics who used to deplore the conformism of modern society would deplore equally loudly the atomized character and lack of community consciousness of the society of their day.

(4) Conservation. Goods and services that constitute income can be classified into two types: those that consume substantial amounts of physical resources and those that do not. Energy, food, and transportation are of the first type. To have more of them one must use up more of the earth's scarce resources. Love, education, and communication are of the second type. They are genuine additions to human welfare, but they do not necessarily use up physical resources, or only incidentally.

However, human beings do not stop congregating as they acquire telecommunications any more than they ceased talking when they learned to write. Communication is not a simple substitute for transportation. The relationship of communication and transportation is two-sided. People do save themselves trips by phoning or writing, but they also develop relationships by communicating, which lead them in turn to travel so as to come together.

While the energy-use consequences of improved communication are not obvious, they are nonetheless profound. The modern city in its present form, with its traffic jams, its high-rise office blocks downtown, and its sprawling suburbs, is a direct consequence of telecommunications developments at the turn of the century. It is easy to believe that the telephone, along with the automobile, was responsible for the dispersion of the population from central cities to suburbia and exurbia. At first glance that might seem obvious, but further investigation reveals how one-sided this assumption is. It turns out that one of the major early effects of the telephone was to make the skyscraper possible. John J. Carty, the chief engineer of AT&T, made the point in 1908. "It may sound ridiculous to say that Bell and his successors were the fathers of . . . the skyscraper. But wait a minute . . . Suppose there was no telephone and every message had to be carried by a personal messenger. How much room do you think the necessary elevators would leave for offices? Such structures would be an economic impossibility?"[17]

Not only did the telephone make the skyscraper possible, but it also helped create the downtown demand for such high-rise buildings. Before the coming of the telephone, business neighborhoods were walking areas. All the traders of a particular kind would congregate in a few blocks, in order to be able to walk up and down the street to do business with one another. Rents in these dense centers of activity were very high. The telephone, the streetcar, and the skyscraper provided an option; they made it possible for firms to escape to cheaper and more commodious quarters. After the 1880s one could escape either by moving *out* or moving *up*.

In other ways, too, the telephone contributed to the growth of downtown business districts. In the nineteenth century at the front of a typical water-powered red-brick factory one would find the office of the president of the firm. He had to be near his plant to control production. With the coming of telephones, the offices were moved downtown, away from the production facility. Orders to the plant manager could be given over the phone with confidence that they would be obeyed. The much subtler and more difficult communication task of negotiating with customers, suppliers, and financiers required the president in person. So corporate offices drew together in the new high-rise office blocks in the city center.

Thus the business phone helped create the dense downtown areas as much as it later spurred the residential flight from them. And as communication changed the patterns of human settlement, it changed the resources we use—the gasoline we consume in our cars, the way we heat our buildings, and the iron, cement, and copper we need.

The computer-based technologies coming along now will probably affect the pattern of human consumption just as profoundly as did the telephone. What the ultimate effect will be is as subtle and complex as the effect of the phone has been. It will become economical to carry on by long-distance communication many activities for which people now cluster together; but the result will not likely be a dispersed population scattered evenly over the face of a rural terrain. It will more likely be a new megalopolitan pattern, where people cluster in new ways, for different reasons, and in different locations.

Examining the impact of changing technologies of communication on society seems to imply a belief in some powerful causality and even determinism. But is technology in command? Certainly it stirs powerful social forces. Whether or not the technologies with which we do our work and conduct our business will control our destinies in the end, they certainly create a massive current. Our question in this book is where that current leads.

The developments described here are happening all over the world, though they have gone further and faster in the United States than anywhere else. Time and space preclude detailed discussion of the workings everywhere, so I slip back and forth, uneasily, from universal comments and multinational examples to a frequent focus on America. While my thesis is global, my illustrations are American more often than I would like.

Chapter 2

The New Communications Technologies

The term *new communications technologies* is shorthand for about 25 main devices, which include cable television, video recorders and discs, satellites, facsimile machines, computer networks, computer information processing, digital switches, optical fibers, lasers, electrostatic reproduction, large-screen and high-definition television, mobile telephones, and new methods of printing. This chapter will provide a brief review of some of the main concepts behind these devices.

Defining a Communication System

Viewed physically, a communication system consists of (1) a series of *nodes*, or terminals, each of which is an *input* device, or an *output* device, or both; (2) a transmission medium among the nodes; (3) sometimes a switching device that determines which nodes are connected to which; and (4) sometimes a *storage* device for holding messages and forwarding them later on. It can be a one-way communication system—like broadcasting—in which one node talks and the rest listen; or it can be a two-way communication system like the telephone.

Capacity

Among the characteristics of a communication system is its capacity—how much information the transmission lines can carry between nodes in a given period of time. One of the intellectual triumphs in this century was the development by Norbert Wiener, Claude Shannon, and others of the mathematics of information. An initial application was to measure the capacity of communication systems. The basic measure is a "bit," that is, the amount of information conveyed by a single binary switch in the on or off position. Any code—numerical, alphabetic, or other—can be represented by some string of binary numbers: zeros or ones. The capacity of a communication system is measured by how many of these bits can be transmitted through it in a period of time.

The lowest bit rate teletype line, known as quarter speed, operates at 12½ bits per second (bps); a telex line or a full-speed teletype line operates at 50 or sometimes 75 bps. An ordinary telephone line can transmit data at between 1,200 bps and 4,800 bps. If conditioned, it may operate at somewhat higher rates, such as 9,600 bps. An uncompressed color television picture uses 75–90 million bits per second (mbs).

The term *bandwidth* represents the same concept as bit rate. The maximum rate with which one can transmit bits electronically is not arbitrary, and there is a theoretical limit for any communication channel, given by a fundamental known as Shannon's law. The practical limits are typically a fraction—between a fifth and a third—of the theoretical limit. A television picture takes about 1,000 times as much bandwidth as a voice phone. Some media like television need a channel only one way; other media need two-way channels. Some devices allow a return channel to be put into a single circuit. All of this slightly complicates the arithmetic.

When we say a particular use "takes" a certain amount of bandwidth, that is shorthand for saying it needs that much bandwidth (or transmission bit rate) to achieve a desired level of fidelity or definition. If we wanted a phone system with better sound reproduction than the present one, it could be obtained at the cost of a system with greater bandwidth. If we want big-screen high-definition television, it takes still greater bandwidth, for a TV picture is just a series of dots, each either on or off. Better definition requires a denser dot pattern.

Thus, whenever we talk about bandwidth or transmission speed, such as bps, we are talking about the capacity a system has or has to have to transmit a given message with a given level of definition.

Digital versus Analog Signals

In any system of electronic transmission the basic principle is that at the input terminal the electromagnetic impulse being transmitted is varied in some way; at the output terminal the original variations in current are reproduced. We have for the most part been describing what is known as *digital* transmission, because capacity measures are best understood in their digital form. In digital transmission the variations in the current or voltage take the form of on-off signals. Another mode of electronic transmission, called *analog*, operates through *continuous* variations in some characteristic of the flow—which may be changes in frequency (FM means frequency modulation) or in amplitude (AM means amplitude modulation).

The telephone, until recently, was an example of an analog device. When Alexander Graham Bell invented the phone, he was seeking to invent a "multiple telegraph." The Morse code was a digital device in conception even if the electronics were analog. The choices were to have the signal off, to have it on for a short pulse (a dot), or on for a long pulse (a dash). The miles (even thousands of miles) of a circuit were completely tied up by a single operator transmitting a single message in that way. The idea occurred to several people, including Bell, that the same wire could carry a number of signals simultaneously if they were transmitted at different frequencies. If one receiver was tuned to one frequency and another receiver tuned to another frequency, then each could receive its message over the same wire, without interference. But while Bell's financial backers were after the economics of a multiple telegraph, what soon preoccupied Bell was a more flexible analog device in which the variations of current mimicked the variations of the frequencies that the human voice made in the air waves. And so the telephone was born.

For a century we have used analog transmission for telephone, but now, in a technically very important development, digital transmission is being increasingly adopted. Digital transmission is not limited to data. It is used for pictures; a telephoto or television picture is reproduced from the original as a sample consisting of thousands of dots. With voice, too, it is possible to sample the frequencies and transmit the sound digitally as thousands of unit frequencies.

Granted it is possible, but why do it? There are substantial economic and technical advantages to digital systems. The recognition and reproduction of a digital signal is more reliable. If a signal has weakened substantially in a long transmission, it may be hard to recognize its exact frequency or amplitude, but whether the bit has sounded at all or not is more certain. A repeater, therefore, can restore the original on-off pattern more exactly. As a result, extremely reliable transmission over long distance with high bit rates can be achieved.

There is another advantage to digital systems. If a message consists of unit pulses instead of continuously varying ups and downs, there are many ways in which one can manipulate the pulses electronically. That is what a computer does. One can, for example, store the bits in a computer memory and then instruct the computer to forward the message later on. Despite the manipulation, the end message sounds or looks just like it would have originally. The digital pulses, or bits, can be transformed to whatever frequency is optimal for transmission or processing. They can be interleaved with each other in "packet networks" (see below) and thus handled more cheaply on a fully loaded circuit than on a circuit devoted to a single connection with all the variations and silences that occur in a single conversation.

In actual systems in use, the separation between analog and digital modes of transmission is not as absolute as described so far. One can sample an analog signal to produce a digital bit stream. Conversely, the beeps of a bit stream can be sent on an analog device which interprets them as simply a discontinuous series of sounds or pulses. Thus, computers or terminals can be connected over an ordinary analog telephone line and chatter to each other in their binary language.

The device that interfaces a digital and analog communication stream is called a *modem*, from *mo*dulator-*dem*odulator. Each of the tonal beeps one hears when one punches in a number on a push-button phone represents some bit pattern in the digital communication code. At the receiving end a modem can interpret the tones and put each out in the original bit code.

Nonelectronic applications

So far we have been following the concepts of electrical engineering; but much of what we have said applies equally to nonelectronic communication. Voice conversation, face-to-face, is an analog transmission by sound waves. The print in this book is digital communication of a stream of fixed unit signals. The fact that there are basically 26 of them instead of two is quite irrelevant; each of the 26 alphabetic characters could equally well be written (as it is in telegraphy or a computer) as a 0/1 pattern instead of in the conventional curvy way we inherited from the Levant.[1]

We can apply the Shannon measure of transmission volume to nonelectrical communication, too. A person talking has a certain capacity, probably about 120 words a minute. It takes a certain number of bits to transmit a word by telegraph, or a different number to transmit it by telephone. One can equate communication volumes sent through nonelectronic and electronic media, either in words or bits as appropriate, depending on the comparison one wants to make. That kind of translation becomes important when we ask such questions as what it would require and cost to convert the postal system to an electronic one, or to deliver newspapers by facsimile. These are questions about how many bits it would take.

Many of the great inventions in modern communications are devices to convert nonelectronic signals to electronic ones or back again. Examples are the telegraph, telephone, microphone, loudspeaker, teletypewriter, facsimile machine, videocamera, television receiver, tape recorder, cathode ray tube, video cassette recorder, and optical disc player. Each starts with an electronic signal and turns it into something humans know how to sense, or vice versa. It is increasingly feasible to convert any given signal to a form in which it can be moved by whatever transmission medium is most convenient and economical.

Transmission Media

Transmission media include those that physically transport the message (like carrier pigeons or letter carriers), those that transmit through the air, and those in which the signal travels in enclosed channels. The last of these have been with us only since the very end of the eighteenth century, when experimentation in electric telegraphy first began. Before that, the limits to which a signal could be sent without travel were the distance that an opaque object could be seen from any given spot on the surface of this curved globe or the distance that an audible sound wave would carry through the air.

Ingenious relays were developed. Smoke signals went from hilltop to hilltop. In 1794 Claude Chappe's semaphore telegraph was installed between Paris and other cities in France. It consisted of a series of towers every few miles whose monstrous mechanical arms were lowered or raised to convey the alphabet by a certain configuration. By this means a message could travel the 525 miles between Paris and Toulon in 20 minutes. But such visual signaling lines were so few and of such low capacity that one can almost say that until the electric telegraph was invented in the nineteenth century, messages never traveled a long distance any faster than a man, animal, or wind-powered craft could carry them. The challenge of finding ways to move messages further, faster, and more cheaply was one of the driving forces in the development of communications technology throughout the nineteenth and twentieth centuries.

Wire networks

The telegraph may have been an obvious invention once it became known that certain metals conducted electricity well, but those metals were expensive. Two of the best conductors, gold and silver, were out of the question. Until about 1877 iron was generally used; then it was recognized that the greater conductivity of copper made it superior for long-distance telephony. The iron wires that had been adequate for telegraphy did not give enough fidelity over long-distance transmission of the larger bandwidth that voice required; the attenuation of the signal was too great.

New mining developments in the last half of the nineteenth century brought the price of copper down just in time for widespread use in telecommunications networks. Nonetheless, the copper itself amounted to a huge part of the total investment in telephone plant before 1900. Improvements in amplifying repeaters, such as the 1907 De Forest vacuum tube and the 1910 Pupin loading coil, made possible effective telephone repeaters. As a result, the American continent was successfully bridged by telephone in 1915.

Radio transmission

Early in this century transmission through enclosed media was improving. Still, inventors from the middle of the nineteenth century on were seeking ways to avoid altogether the enormous expense of the wire network, by transmitting signals unenclosed. They understood that the electric circuit could be completed by ground return, and many early telephone and telegraph systems were built with one wire only. Scientists also knew about electromagnetic induction (the principle behind the electric motor and the electric generator), so they knew that electricity could carry impulses through the air. And, of course, there was light. As early as 1879 Alexander Graham Bell had developed a selenium photophone to transmit speech by light instead of electricity.

In 1890 the German physicist Heinrich Hertz identified what we now call radio waves. Given the successful transmissions that had been achieved with electricity and light, it occurred to many technical people that these hertzian waves could also be used for signaling. The practical issues concerned what kind of device it would take to modulate and detect such waves and how far they would travel. Guglielmo Marconi believed that hertzian waves would travel a very long way. On December 1, 1901, he succeeded in a sensational experiment of transmitting telegraphic code by radio across the Atlantic from Poldu, Wales, to Newfoundland. Five years later Reginald Fessenden managed to transmit distinguishable spoken Christmas Eve greetings by radio to ships at sea. Signals good enough for practical voice applications had to await a better method of amplification, however.

This was made possible by Lee De Forest's vacuum tube, which came on the scene in 1907, followed in 1913 by Edwin Howard Armstrong's regenerative feedback circuit. These and other achievements raised the question whether telegraphic and telephonic communication would continue to go over wires at all, or whether the economy of wireless transmission would sweep the day. With hindsight we know that it did not. One major reason is that if many signals go forth into the air, they are apt to interfere with each other, because of the limitations of spectrum.

To deal with the problem of interference, in 1906 the International Telecommunication Union (ITU) took on the task of registering radio frequencies in use so others would not infringe on them. After the U.S. Navy complained that radio amateurs were interfering with its transmissions to ships at sea, Congress, in 1912, established a system of licensing radio transmitters. Five years after radio broadcasting began, in 1926, the courts threw out the 1912 Act as arbitrary and unconstitutional.[2] Interference by broadcasters with one another quickly became so severe that the industry itself appealed to the U.S. government to set up some system of licensing so that radio stations could be alone in their segment of the spectrum.

Coaxial cable

For these and other reasons, there were many predictions in the early years of radio that entertainment broadcasting would be delivered by wire rather than over the air. In the end, however, the cheapness of delivering a uniform message to millions without the capital plant required to reach each of them by wire made over-the-air broadcasting succeed. In an advertisement-based system in which the advertiser will pay only a very small amount per person exposed, no other method would have been as practical.

There has always been a trade-off between the advantages of delivery enclosed within insulation, which is generally more expensive but theoretically unlimited in its capacity and largely free from major interference, and the advantages of over-the-air delivery, which is generally cheap, potentially oligopolistic because of bandwidth scarcity, and subject to interference and even public health limitations.

The pendulum continues to swing. As mentioned, from the time the phone was invented to the 1920s, many observers forecasted radio-like broadcasting on the telephone; and early on there was speculation about wired television.[3] Once broadcasting was established over-the-air, to those who do not know that history, cable television appeared as a new thing. And in some ways it is. Taking advantage of the great bandwidth capacity of coaxial cable (a development of the late 1930s) in place of twisted wire pairs, cable TV offers the economy of delivering several dozen channels of video on a single cable. With regular wires, the only way to come close to offering so many alternative channels is with the added expense of having a switching exchange so the viewer can select for transmission over his wire the one television signal he wants. (Even then, regular telephone wires do not have enough bandwidth.) That is known as *rediffusion* TV, or video on demand. But the cost of switching is not trivial, and rediffusion cable systems are accordingly more expensive for carrying many channels than coaxial cable. The attraction of moving TV activity to cable is high. One resistance is the reluctance of customers to start paying for programs they have been receiving free. Nonetheless, cable TV has grown tremendously. In the United States the majority of households now subscribe to cable, and the number increases each year.

Optical fibers

Within a few years, the pendulum is likely to swing further, owing to the progress made in developing a new enclosed transmission medium, *optical fibers*. Light does not escape out the sides of the fiber because its walls act like a mirror, reflecting the light back in. Thus, the light is conducted within the fiber to the end even if the conduit is curved. With optical fibers, Bell's notion of 1879 is being realized; modulation of light is used for signaling, but not through the air as Bell expected, where other beams of light may interfere.

Optical fibers were a familiar technology before 1969, but the attenuation of the signal over distance was severe (the same problem long-distance telephony had before the Pupin coil). In 1968 glass fibers were developed that were of sufficient purity to allow transmission of up to a kilometer before reamplification. Immediately, a race was on among America, Europe, and Japan to purify the fibers, attach their ends together, switch them in a network, and develop amplifiers and even switches for use with them. The attraction was economy. Thin glass fibers by the thousands would fit into underground ducts now often filled to capacity with thick coaxial cables. Even more important in the long run, glass is made of cheap silicon—sand—found all over the world. Copper, like oil, is approaching depletion. Best of all, the obtainable bandwidth of optical fibers is enormous and is continuously expanding.

Another reason for restricting use of over-the-air transmission is the growing demand for use of spectrum in situations where insulated media are impractical. Mobile vehicles and satellites cannot be wired, and more spectrum is being required for this increasing traffic. Thus, the prospect is for greater use of insulated media in those situations where they are a viable option.

Spectrum

A characteristic by which one classifies radio waves is their *wavelength* or *frequency*—two terms describing the same thing. Long waves oscillate at low frequency, short waves at a higher frequency. All electromagnetic signals travel close to the speed of light, regardless of their frequency. Thus, a measure of frequency of wave peaks passing a given point is the inverse of the length of the wave.

Transmissions at different wavelengths have different characteristics, and these determine the uses to which they are put. Marconi found that sending telegraph signals over the great distance between continents required long waves, which follow the curvature of the earth. But to efficiently send or receive signals with such long waves required large antennas. Thus, in the first part of the century "antenna farms" spread out near the shores.

When broadcasting came along in the 1920s, broadcasters used somewhat shorter wavelengths, called medium waves (300 to 3,000 KHz). With these the receiving antenna could be much smaller. On the other hand, reception of broadcast stations would only reach a few hundred miles at most. The next step in the 1930s was to short waves. Short waves, however, do not follow the curvature of the earth as long waves do but, like light, go primarily in straight lines. Their direct reception is thus limited to line of sight, or in other words to a range of perhaps 30 or so miles radius around the antenna tower. Fortunately, radio amateurs quickly discovered that short-wave frequencies are reflected back to earth by the ionosphere. So short waves, like long waves, were found to be useful for long-distance broadcasting, though by a mechanism of reflection rather than bending.

The exhaustion of the spectrum by further claims on frequencies was postponed as engineers learned how to squeeze more stations into the same frequency range without interference. But even more important, they gradually mastered the use of shorter and shorter waves. When FM radio and television were established in the 1930s, they were assigned to wavelengths called very high frequency or VHF (30 to 300 MHz). When still more television channels were needed after the war, they were assigned to an even higher band, ultra high frequency or UHF (300 to 3,000 MHz) that had been made accessible by radar research. UHF has slightly less desirable characteristics than VHF, such as a shorter range; but no more VHF spectrum was readily available. High frequencies were also used by the expanding network of telecommunication relays. Microwave transmission, going from tower to tower at line-of-sight distances and capable of being focused into narrow beams, soon carried more telephone long-line circuits than did coaxial cables. Today, some transmissions take place in the giga-hertz band (billions of cycles per second), where wavelengths measure in centimeters, in contrast with the hundreds of meters of medium and long-wave frequencies.

Newer services for which spectrum allocations are needed include satellite transmissions, mobile radio to cars, airplanes, and trucks, and citizen band radio. Spectrum is still available (without displacing current users) at high frequencies above 14 and 16 gigahertz and particularly in the still-empty portions of the spectrum from 40 to 300 GHz. But such high frequencies—whose character begins to approach that of visible light, which is at still higher frequencies—do not transmit well through rain and snow clouds, and the receiving equipment has to be more sophisticated and expensive. Many users would like to use lower frequencies, just as broadcasters prefer VHF to UHF.

One might wonder why frequencies cannot be simply reassigned so as to be used rationally and optimally. The answer is that with spectrum, as with land, those who have prior possession are not easily moved. Billions of dollars have been invested in equipment designed for certain frequencies. Reassigning frequencies makes that equipment obsolete. That is the obstacle, for example, to correcting the decision made some years ago to have television stations operate in two separate but unequal parts of the spectrum, VHF and UHF.

Satellites

Readers of science fiction have known for many decades that if a rocket achieves a certain velocity it will escape from the earth's gravity. At a lower velocity it will neither escape nor return to earth but will circle endlessly, like a moon. And at still lower velocities it will be drawn back by gravity and eventually crash to the earth's surface. At one particular height and velocity, an orbiting body will circle the earth once a day, and thus, if it is placed in an orbit that follows the equator, it will remain always over the same spot above the rotating surface of the earth. From the point of view of someone on the earth, such a satellite, called a *geosynchronous satellite,* would look stationary.

There is only one geosynchronous orbit. It is over the equator, because an object in nonequatorial orbit around the circumference of the world would seem from earth to move north and south as it cycled around the world. Furthermore, a geosynchronous satellite must be at a particular altitude, namely 36,000 kilometers or 22,300 miles, in order to maintain an orbiting speed that exactly matches the speed of earth's rotation. At such an orbit satellites seem to stand still. For communication, that is a very convenient property. If one wants to set up an antenna on earth and beam a signal to a satellite or receive a signal from a satellite, it is much cheaper and simpler if one can point the antenna once and for all, rather than having to track the satellite with a moving antenna controlled by complex works.

There are disadvantages to the geosynchronous orbit. Since it is over the equator, the North and South Poles are blind spots, and communication to very high latitudes is not very good. From high latitudes it is also hard to launch a satellite into orbit around the equator. Between 1963, when the United States launched Syncom 2, and 1974, when the Soviets successfully placed Stationar I, the United States was the only nation able to put a satellite in geosynchronous orbit. Up until 1974 the Soviet Union had relied exclusively on nonstationary Orbita satellites for communication.

Another problem with the geosynchronous orbit is its limited size. If satellites are too closely spaced, there will be signal or even physical interference between them. The spacing required depends upon the power of the signal, the precision of the beam, the quality of the ground receiving antennas, and various other technical features such as polarization and frequency.

How serious is the crowding problem? With one-degree separation there could be 360 satellites around the world using the same or related frequencies. That sounds adequate, but the number is deceptive. Different locations on the orbit are not of equal value. There are long stretches over the Pacific and Indian oceans where demand for satellite communication is low. Spots useful for the countries of Europe and America, on the other hand, are scarce and very much in demand. The Americas lie on a north–south axis, which limits the degrees of arc in the geosynchronous orbit that can service the Western Hemisphere. And the same few degrees of longitude are of interest to all of the approximately 30 countries of the European continent. If every country ended up wanting its own geosynchronous satellite, there would be a serious crowding problem.

The shortage of geosynchronous arc requires international cooperation. But some nations launch satellites for reasons of national pride and national security. If there are to be national satellites, and if they can interfere with one another, then some control and allocation mechanism has to be worked out.

The system in use for assignment of spectrum has begun to be extended to satellite locations. At the World Administrative Radio Conferences (WARCs), frequency bands are allocated to different uses such as broadcasting, satellites, and marine communication. When a nation plans to use a particular frequency within a band, it registers the frequency with the International Frequency Registration Board of the International Telecommunication Union (ITU). Any other nation may protest that the signal will interfere with some frequency of its own. The requirement is only to consult; there is no compulsion mechanism, but bad behavior may cause reciprocated nastiness, so countries try to work things out. Since the 1979 Space WARC, the system for assigning frequencies has been extended to orbital locations. Orbital slots are allocated on a flexible, regional basis, in order to give each nation access to a slot while also allowing for growth. Nations register their satellite launches, locations, and frequencies. Other nations may protest.

A pair of frequencies is required for satellite communication, one for uplink signals (from the ground to the satellite) and a different one for downlink signals (from the satellite to the ground). They must be different to avoid interference at the satellite. Consequently, on the satellite are one or more transponders, devices that receive a radio signal at one frequency and retransmit it at another. The satellite is, in effect, a microwave relay station high in the sky.

The design of a satellite system may vary on a number of parameters, such as frequencies used, the direction and width of beam, and power. Transmissions from high-powered, sophisticated, and therefore expensive satellites can be received by small, low-cost ground stations. If a weaker signal is sent from space, more ex-

pensive equipment on the ground is required to receive it. Sophisticated satellites, in addition to having high power, may have
storage and switching capacity and the ability to vary the pointing
and focus of the beams. At any given power level, the narrower
the beam, the stronger the signal that will be received on the
ground. The early satellites sent their weak signals in a global
beam covering approximately one third of the earth's surface.
Since that is the maximum area a single satellite can effectively
cover, it takes only three satellites for a global system.

The choice between investing more in the space segment or
more in the ground segment of a system is a matter of economics
and purpose. If, as in the case of INTELSAT (the international satellite communications organization), the main purpose is to provide
intercontinental links among national telecommunications systems, then relatively few ground stations are needed, typically
one to a country. In that case it pays to make that one earth station
large and expensive and save money on the still more expensive
satellite system up in space. On the other hand, if the purpose is
to provide television directly to Indian villages, of which there are
500,000, then it is critical to make the half million ground stations
as cheaply as possible, even if the satellite has to be more expensive.

Computer Message Processing

There was a time when electronic communication and computation were thought of as quite separate and distinct activities. Today they are intertwined to the point where no meaningful borderline between them can be drawn. Both a computer and a digital
telecommunication system can be described as devices that switch
bits of information around under the control of a stored program.
Each has a memory in which signals are stored. Each accepts input
signals from terminals and each also sends output signals to other
terminals.

In a typical computer of the 1960s the transmission path from
the terminal to the input or output buffer (a temporary memory)
was probably a few hundred feet; from that to the central processing unit (CPU), where switching was done, was a few feet. The
output terminal was in turn a few hundred feet from the CPU,
usually located quite close to the input. But by the 1970s many

computer systems had become widespread networks of dispersed processors and memories. The terminals may be the cash registers in stores or airline reservation desks. These are part of a time-sharing system scattered over hundreds or thousands of miles. The results of each operation may be recorded in memories located in different places and processed in other places among a network of components.

A particularly significant kind of data transmission system is one called a *packet network*. This name comes from the fact that the messages, when sent, are divided up into uniform-sized packets of bits—often 1,288-bit characters. Each packet has a header noting the message from which it is drawn, its position in the message, and the destination to which it is addressed. The nodes that the network connects may be terminals or computers. Between each node and the network itself sits a small interface computer that breaks the messages up into packets and sends each off to its destination individually but interleaved with packets from other messages. Each packet is sent forward on any route that has room. The interface computer also receives and stores incoming packets and passes on, or switches, packets destined for other nodes. This allows high-speed lines to be used to capacity, since the line is not dedicated to a single slow typist or speaker. When the incoming packets from a message are all in, the interface computer at the destination reconstructs the message and passes it on to the terminal or computer to which it is addressed. Despite these complex operations, end-to-end transmission times are very short, typically 100 milliseconds for distances up to several thousand miles.

Having reviewed many of the technical terms and concepts used in this book, we can now turn to the social implications of these developments in modern electronic communication.

Chapter 3

Crumbling Walls of Distance

Pick up your telephone; after 11:00 o'clock at night it will cost you only pennies for the first minute to call from coast to coast in the United States. To call from New York to Chicago, one third of the distance, will cost almost the same. In the jargon of telecommunications, these rates are *distance insensitive*.

Evidence of distance insensitivity in telecommunications is all around us. A movie shown in New Zealand is likely to have been made in the United States or Europe; it costs the same to see it in New Zealand as where it was made, since the charge for shipping the can of film is the smallest part of the total cost. A television show in Thailand is likely to have been made abroad, but that does not make it expensive in Thailand, again because transit is a minor item in the cost. A newspaper in the United States has news of what happened in the past 36 hours around the world. In choosing what to carry, the American editor considers the degree of interest of his readers in what happened far away, but not the extra cost of long-distance transmission; the news from everywhere has been sent by the news service at a flat fee. Put your own message in the mail and the postage will be the same whether the letter is delivered locally or across the country. The real cost is, of course, not exactly the same, but there is surprisingly little difference; the long-distance transportation of the letter is a small part of the cost of handling it.

This has not always been so. When messages were carried by men or animals, transporting a message to three or six thousand miles was a very expensive task. Distance was a major barrier to interaction.

Instantaneous Communication

On the day in 1858 when President James Buchanan and Queen Victoria exchanged greetings through the first cable to cross the Atlantic, the map of the world changed. Until that moment nothing that happened in either hemisphere could be known in the other in less than the two weeks or more that it took a ship to cross the ocean. Since then (or, more accurately, since the day eight years later when a permanently working cable was finally installed), Europe and America have had instant news of each other.

In the Battle of New Orleans in 1814 over 2,000 soldiers died in what appeared to be the decisive battle of a war between Great Britain and the United States. But unknown to those in the battle, the war had ended by treaty two weeks before. In both politics and trade such contretemps were normal before telegraphy. No wonder that the miracle of instantaneous communication caught the Victorian imagination. In a typical "pop" expression, past plagiarized from Tennyson, a contemporary versifier wrote

> Lo the golden age is come,
> Light has broken o'er the world.
> Let the cannon mouth be dumb,
> Let the battle flag be furled;
> God has sent me to the nations
> To unite them, that each man
> Of all future generations
> May be cosmopolitan.[1]

Conquering the Atlantic was the most dramatic leap in the compression of distance, but the process had been under way since the birth of the telegraph. Most early telegraphs were too cumbersome, expensive, and unreliable for practical use. Samuel Morse's system was simple and rugged. Skilled human operators had to interpret the coded output of dots and dashes. So for short but important messages, like flash news or business agreements, instantaneous communication over distance was being established by the middle of the nineteenth century, though by a labor-intensive means and at high cost.

The telephone, when it was invented in 1876, was perceived by contemporaries as a money-saving variant of the telegraph. Even after Bell's backers knew they had a voice telephone and not just a telegraph for multiple frequencies, they sought short-term profit from their invention by offering the patent for $100,000 to Western Union, which had wires already in place. If the telephone was a sophisticated telegraph, Western Union seemed the natural client for it. Shortsightedly, Western Union declined the offer.

Initially, putting phones at the remote ends of long-distance telegraph lines was not feasible. The original telephones operated only over a maximum of about 20 miles. But the telephone entrepreneurs believed correctly that the technical obstacles for long-distance telephony would be overcome. Sir William Thomson, later Lord Kelvin, at the first public exposure of the telephone at the Philadelphia Centennial Exhibition of 1876, assumed in his laudatory comments that the device would be used for long-distance talk. Bell, in a letter in 1878, declared: "I believe in the future wires will unite the head offices of the Telephone Company in different cities, and a man in one part of the country may communicate by word of mouth with another in a different place."[2] Theodore Vail, the President and builder of the Bell System, wrote his staff in 1879: "Tell our agents that we have a proposition on foot to connect the different cities for the purpose of personal communication."[3] He expressed himself as confident that "Mr. Bell will give us the means of making voice and spoken words audible through the electric wire to an ear hundreds of miles distant." The 1885 charter of AT&T said that: "The general route lines of this association . . . will connect one or more points in each and every city, town or place in the State of New York with . . . each and every other city, town or place in said state, and in each and every other of the United States, and in Canada and Mexico; . . . and also by cable and other appropriate means with the rest of the known world."[4]

The first long-distance line, in 1881, connected Boston to Providence, and the second linked Boston with New York. By 1892 a line from New York reached Chicago. All these early conduits had enormous technical problems, especially attenuation with distance. Understanding voice required much higher fidelity, and therefore more power, than did dots and dashes.

There were various ways to maintain the quality of the signal. One was to increase the quality and thickness of the wire; a second was to raise the strength of the current; a third was the installation of amplifiers and repeaters along the line to build up and regenerate the signal. The first technical improvement came when copper wire was substituted for iron wire on the Boston–New York line. But solving the problems of long-distance telephony by improving the wire or increasing the current was expensive. The New York–Chicago line consumed 870,000 pounds of copper wire. Before 1900 and Pupin's invention of the loading coil, about one fourth of all the capital invested in the phone system was spent on copper. After the introduction of the loading coil, the diameter of the wires could be cut in half. Nevertheless, to make a transcontinental phone line economically feasible, high-quality repeaters were necessary. In the period around the turn of the century the most important objective of telephone researchers was progress on repeaters. This was accomplished with the perfection of the vacuum tube, and in 1915 a transcontinental line was completed.

By 1919 the main component of the cost of a long-distance call was the investment in wires and in the attached repeaters, poles, and similar equipment. Those costs were roughly proportional to the distance traveled. A call that went twice as far needed twice as many miles of wire. But as transmission technology improved, transmission itself became a smaller and smaller part of the total cost of a long-distance call. Billing, switching, and the local loop at each end became more significant elements of the total cost, and later the dominant portion of it. These costs are the same for each call, regardless of distance. Thus, even without radio and before the advent of satellites, long-distance transmission costs were becoming a minor part of the total bill. AT&T thinks nothing of routing calls intended to go up and down one coast of the United States through connections that cross the country from coast to coast and back again if direct circuits are busy or incapacitated. Load leveling is far more important in cutting cost than is minimizing distance. The marginal cost of extra miles of transmission is a vanishing number, and even the average cost is just pennies for such distances as across the United States or across an ocean.

Wireless Transmission

The need to communicate with places where wires could not be strung at all, as to ships at sea, or where the cost of wires was a heavy burden led to an interest in a wireless telephone. The first effort at a wireless phone (which absorbed Bell's efforts in 1879–80) attempted to use light as a transmission medium. However, until waveguides and optical fibers provided an entrapped light channel, interference in the natural environment precluded displacement of wire transmission by light.

In April 1901, only nine months before Marconi's radio transmission from Poldhu, Wales, to Newfoundland but well after his 1896 patents, General John Carty (later the chief engineer of AT&T), in his anonymous "Prophet's Column" in the *Electrical Review*, evaluated radio waves versus light as a possible means for long-distance transmission and concluded that the prospects for light were better in the short term.[5] But Carty underestimated the imminence of radio communication. After Marconi's demonstration, interest turned rapidly from light to the invisible hertzian waves.

Marconi was not alone in experimenting on hertzian waves for message transmission. What distinguished Marconi was a deep faith, which turned out to be correct, that radio waves could travel very far and still be successfully detected. Like Morse, he was a promoter; he caught the world's imagination by his transatlantic feat.

The drawback of radio transmission was interference. There was simply not enough spectrum to allow unlimited use of it by all who might wish to communicate. The problem of interference was one determinant for the telephone system's staying on the wire network except for special purposes.

For one-to-many broadcasting, however, where the same few messages were addressed to large numbers of receivers, over-the-air radio was an economic godsend. With radio, one did not need a wire circuit to each member of the audience. When used for this purpose, radio transmission eroded distance as a barrier to communication.

Next came communications satellites. Here, it makes little difference whether the two points on the ground are 500 or 5,000 miles apart. If distance was already a minor cost factor before, with satellites it became trivial.[6]

Various studies have been made of the distance of the crossover point beyond which providing satellite transmission becomes cheaper than using a terrestrial net. The exact distance depends on many details of the design of the system. Estimates of the crossover range as low as 300 miles.[7] The basic conclusion is clear: distance has ceased being an important factor in communication costs.

Unless national policies bar it, the rates for long-distance tele-communications will eventually fall so low that volume will grow enormously. Rates, of course, often lag behind costs. Moreover, political pressures favor subsidy to home users of local service at the expense of the institutional users of long-distance communications. Most post and telecommunications administrations are inclined to make as much profit as they can on international business in order to offset losses on the politically protected basic residential service.

But if either the United States or Canada, for example, were to set unreasonable long-distance rates on their side of the border, it might pay to detour data across the border and back. Pricing of some telecommunications services above cost becomes hard to maintain in a world where foreign services can be "imported" at little extra costs over those operating from close at hand.

I have been discussing here just one element of the cost of communication, namely long-haul costs. That is the domain in which communication costs have fallen most and will continue to fall. The costs of switching, billing, and the local loop are by now the greatest part of the costs of a long-distance phone call or telex. Human handling of transcription and delivery are the major items in the cost of an old-fashioned telegram. The cost of programs and receivers are the bulk of the cost of broadcasting. Those factors do not vary with distance. A rise in those costs could offset the savings that result from the cheapness of long-haul transmission. So my point is not that communication is necessarily becoming cheap. Rather, the point has to do specifically with the *geographic pattern* of communication. Until now, the fact that interacting over a distance costs very much more than interacting with those close at hand meant that we lived our lives in neighborhoods. The surface of the world was partitioned among communities, and most activity was within these confines. For almost any activity, one could draw a curve of frequency of interactions by distance, and the shape of the curve would show a rapid falling off with distance. Increasingly now at least one of the causal factors behind that curve—namely, cost of long-distance communication—is vanishing.

Chapter 4

Limits to Growth

Along with the erosion of barriers of distance and the convergence of different modes of communication into a single electronic system, a third major trend in the evolution of communication that deserves our attention is the increase in the sheer quantity of messages transmitted. Observers talk of an information explosion, of information overload, and of an information society.

For thousands of years of human history only a tiny minority of mankind has engaged in intellective activities professionally; at least in the most advanced countries such activities are now the norm of the majority. The rapid and continued growth of this information sector is partly a measure of how important it is and partly a measure of how inefficient its contributions have been in the past. The decline of farmers from more than half of our work force to 4 percent does not mean that food is any less important to us than before. It reflects the extraordinary success of applying technology to achieving efficiency in an activity that absorbed so much of human effort previously.

Spectrum: A Finite Resource?

As mentioned in Chapter 2, one pressing limit to growth of communications is the electromagnetic spectrum. In the United States there are thousands of radio and television stations, over 15 million transmitters on citizens' band, hundreds of thousands of radio amateurs, and transmitters in ships, airplanes, taxis, trucks, and police cars. There are cellular telephones, cordless telephones, microwave circuits, and transmission to and from satellites. And on top of all of that, about half the spectrum has been pre-empted by the government, much of it for national security use. How can anyone judge the relative value of such alternative uses? How can one choose between various applications and applicants, each of whom is staking a claim for the same resource? The government, as an applicant, argues the primacy of national defense. The cellular service users argue that they cannot string wires to their audience as an alternative way of carrying signals. The broadcasters claim that the entire public is served by them, not just special interests.

The usable spectrum has been expanded by technology, as we have learned to use higher and higher frequencies and also to pack more and more information into the same bandwidth. Yet there is a theoretical limit beyond which further expansion will simply not be feasible, particularly as the higher frequencies are in some respects less efficient and economical. It is like drilling in poorer and more remote oil fields as the supply runs down.

As that limit is approached, and as growing communication services require ever more bandwidth, society is forced to make choices. Some claimants will get spectrum for over-the-air uses; others will not. For those who do not have access to the "free" airwaves, there is an alternative: they can send their signals through enclosed media such as wires, cables, waveguides, and optical fibers, insulated from interference with the outside.

Along with the disadvantage of having to pay higher prices, there is a positive side to moving off the airwaves. With coaxial cable and optical fibers, it is possible to use an amount of bandwidth that no one ever had available over the air. One can, for example, transmit 70 or even more television channels over a single coaxial cable. Instead of having, say, four TV stations in an average city, it becomes possible to have many dozens; if one uses multiple cables or fibers, there are no theoretical limits to what the Sloan Commission has called the television of abundance.[1]

Since enclosed transmission media are expensive, much of the art of electrical engineers has gone into finding ways to bring their costs down, either by using less costly equipment or by cramming more traffic onto them. Since Bell's attempt to devise a multiple telegraph that would send several messages over a single wire, we have steadily learned to pack more and more messages into a single carrier. Besides dividing up the frequency band into subdivisions, it is also possible to split up the time on a given channel (time division).

But the main way to save on costs in an enclosed transmission medium is to find a cheap material with a large bandwidth. Today, the biggest economy is from optical fibers constructed of glass. A single glass fiber no thicker than a human hair has the bandwidth of several video channels. In the long run, the important promise of optical fibers is this extraordinary bandwidth. If a fiber can carry, for example, many gigabits of transmission and if that fiber is cheap enough to replace the copper-wire telephone drops to people's houses, then all the various services that now come to the home—over the air, over telephone lines, and over coaxial cable—could perhaps all be carried on the same drop. This raises important policy questions, for example about a potential re-monopolization of communications.

Spectrum markets

Spectrum is not so much a scarce resource as it is a misused one. The main reason for the misuse of spectrum is that it is usually given away for free. Nothing is more subversive of social equality than for the state to distribute a limited resource as a privilege. Whenever resources are allocated that way, they go to possessors of political skill or power who use the resource wastefully, leaving little for the use of others. That was the character of feudalism; the monarchs' favorites got hunting reserves and prime agricultural land; the "outs" had to make do with marginal plots. That is the way that spectrum is allocated wherever its use is licensed. This does not imply malfeasance—feudal monarchs could, after all, be benevolent too.

In the United States the applicants for a television broadcasting license present briefs to the Federal Communications Commission (FCC) describing how they will use the privilege of exclusive use of 6 MHz of bandwidth. After hearings, some applicants get the bandwidth if the Commission is convinced that they will serve the needs of the community better than others who are competing for the privilege.

The FCC applies its own social values to judging what the applicant says about programming for public affairs, religion, and children, how representative the applicant's organization is of the community, and how responsible it is. To the winner in this contest of political judgment goes control of what is often one third or one fourth of the TV spectrum allocation available in the community. Applying the same kind of political judgment, the U.S. government has kept about half the spectrum for its own use.

The fact that spectrum is given initially at no cost to those who are favored, while they have to buy other resources, inevitably distorts the way spectrum and other resources are used. Any rational producer will use factors of production in accordance with their price. If there is a free resource, he will use as much of it as he possibly can in replacement of things for which he would have to spend money. If a river passes his plant, he will use that water to the maximum rather than buying fuel or carting away waste. If no one else is deprived, so be it; but if others lose use of the water because of contamination, then a charge on the water may make him consider alternative methods of waste disposal.

In general, those who have licenses to use spectrum do so with deplorable inefficiency because it would cost them something to use it more carefully. Let us look at two examples of how they would behave differently if they were charged a fee, and how that would leave more spectrum for others. One example is land-mobile radios; these are the radios that are found in taxis, trucks, and other vehicles. In 1974 Robert Crandall of the Brookings Institute and John Ward of the Massachusetts Institute of Technology were asked to look at the growing congestion in the land-mobile portion of the spectrum. The number of licensees had increased exponentially, and the industry was asking the FCC to reallocate more channels to it, from as yet unused allocations. In busy places like Chicago, users were complaining that interference was degrading service.

Land-mobile communication is a valuable service that increases productivity. Taxicabs can keep full a larger portion of their time and waste less gasoline in cruising. Delivery trucks can be routed more efficiently if they keep in touch with a central dispatcher. Garbage removal can be made more economical if the trucks can report as they begin to fill up to capacity. With no charge for radio spectrum, the land-mobile users buy low-cost, low-quality two-way radios. But if the assignments were costly, or for narrower bands, or at higher frequency, or with lower power limits, the users would equip their vehicles with a more expensive and sophisticated transceiver. The federal government thus subsidizes the vehicle owners by giving them much-wanted portions of the spectrum free of charge, thereby allowing them to substitute profligate use of the spectrum for investment in better equipment.

Crandall and Ward concluded that if the users had to pay even a modest charge for the spectrum they got, provided that charge was relative to the amount and quality of the spectrum they wanted, there would be no congestion whatever. Many licensees whose needs were small but who had installed the systems because they were cheap would drop them entirely. Others, noting that their charges for spectrum could be lowered if they bought better two-way radios, would find that it paid them to do so, and they would use spectrum more efficiently. But why invest in better equipment if the bandwidth for voice is free?

Broadcasting uses the airwaves with archaic inefficiency. One reason is that in 1921 AT&T lost to the commercial broadcasters in the struggle for control of broadcasting stations.[2] Consider what would have happened if AT&T had been more aggressive in radio research in the first decade of the century and had dominated RCA's patent position when the confrontation came. It might then have been able to implement its broadcasting plan to operate as a common carrier for all who wanted to buy time to broadcast. At least two possible paths of development suggest themselves. One possibility is that the government would not have tolerated AT&T's growing monopoly position and either nationalized the system (giving the United States a state-run broadcasting system as in Europe) or broken it up. But suppose the government had allowed AT&T to carry out its plan. There would probably be a broadcasting system quite unlike any that exists in the world today. It would, among other things, have been more parsimonious with spectrum and thus technically more advanced.

Today, phone systems are usually dominated by engineers, but in broadcasting organizations engineers play a rather lowly role. The top positions in commercial systems are usually held by people from the programming or marketing side. The AT&T plan would, however, have made economical transmission the key to success in the business. Since revenue would depend on leasing as much air time as possible, the broadcasters would probably have developed a system with more efficient use of available spectrum so as to maximize the number of broadcast channels that could be rented; they would also probably have developed a pricing scheme that allowed for market segmentation so that special-interest broadcasters could buy air time without unduly lowering the rates simultaneously charged to mass broadcasters. If broadcast transmission were placed under rate-of-return regulation, there would be all the more incentive to invest large amounts in transmission plant to create a high-quality and very efficient broadcasting plant, and the engineers in charge certainly could have advanced the art way beyond what has been achieved today. On the other hand, the extraordinary development of American entertainment programming would probably have been less advanced. The broadcast industry achieved well in the area it cared about—the ratings game. It did not perform in an area that did not matter to it—technically efficient use of the spectrum.[3]

The proposal to create markets in spectrum has always been popular with economists, who recognize how much it could improve allocations. But it has always been rejected by the broadcast industry and by those public officials who allocate the spectrum. Those who operate the present system do not believe that the blind forces of a market could possibly take account of all the complex considerations they now weigh. The people who raise such objections, not being economists, only dimly realize that markets are not necessarily all alike; one does not have to sell spectrum in the way that one sells soap. Nor is it an all-or-nothing issue; market structures can take account of natural monopolies and public goods that require political regulation. The spectrum market would most resemble the real estate market with its zoning, rights of way, rights of neighbors, and building codes. Yet a real estate market gives consumers far more freedom than a system in which housing is built and allocated by the state.

All these considerations have been superbly laid out by Charles Jackson. He proposes several different market structures for different uses of spectrum. The situation in which there is plenty of spectrum for all who want some but in which there are already problems of interference (as was the case in 1906 when the Berlin conference first set standards for radio) is different from a situation in which there are few desirable frequencies left to assign. Situations in which channels can be shared, as on CB radio, are different from those in which allocations are exclusive, as in entertainment broadcasting. In general, Jackson concludes, "Spectrum

scarcity clouds men's minds. It is a phantom."[4] It is just a consequence of failing to place an appropriate charge on a limited resource.

Jackson also shows that electronic hardware and the spectrum
are complementary resources. It is hard to maintain both regulated allocation of spectrum and a free market in hardware; those
are two wedded parts of the same activity. If regulators face a
shortage of spectrum (as for land-mobile services) or have available a mixture of more desirable and less desirable frequencies (as
in VHF and UHF television), they are tempted to regulate the sets
on the market so as to solve the spectrum problem. Congress did
that in 1963 when it legislated that all TV sets must be able to
receive UHF as well as VHF. Conversely if one wishes to keep the
hardware industry in the domain of the free market, one ought to
favor a market for spectrum, too.

An abundant resource

If I seem to be arguing both sides of the question of spectrum
scarcity—that society will be forced to more expensive means of
getting spectrum and, at the same time, that there is no shortage
of it—it is because the discussion of spectrum is so replete with
opposite errors. There are extreme and contradictory clichés that
need rebuttal.

On the one hand, there is the proposition at the foundation of
all radio and television law in the United States that the spectrum
is a uniquely scarce resource and that this justifies a regulatory
regime at variance with all the traditions of the First Amendment.
"Unlike other modes of expression, radio inherently is not available to all. That is its unique characteristic, and that is why, unlike
other modes of expression, it is subject to government regulation."[5] That proposition, we have been arguing, is simply not so.
There is a great deal of spectrum, just as there is a great deal of
woodpulp for paper or of assembly halls for meetings, but of
course not an unlimited amount. The limited supplies of each can
be squandered or used well. They can be rationed out by a market
or rationed out in a political way by the authoritarian practices of
government.

A contrary proposition, equally false, is that we are about to
escape from any shortage of spectrum by the magic of electronics.

Blue-sky speculators suggest that we are about to move from an era of communications scarcity to an era of communications abundance. The oligopoly of television broadcasting will give way to the open channels of cable TV. Multimillion dollar professional studios will be challenged by adolescents with camcorders. The abundant bandwidth of optical fibers or the 24 transponders on a single satellite, each pouring out millions of bits a second, will provide, it is argued, all the bandwidth that anyone will need in order to converse freely, by voice and pictures, with all the world. All of these forecasts are individually true, but the generalization that spectrum will be no problem does not follow. While relative to the present the forecast of abundance of bandwidth is probably valid, in any absolute sense it is false. We are more likely to be moving from an era of scarcity not to an era of abundance but to another era of scarcity at a higher level.

As with the fantasy of "an economy of abundance," the fantasy of communications abundance is defeated by man's extraordinary capacity to augment his wants. Keep in mind the difference in bandwidth requirements for text, voice, and pictures. Text can be represented by 6 to 9 bits a letter. A telephone-quality voice-grade circuit operates satisfactorily at 1,200 bits a second, or can carry 9,600 bits per second and more if conditioned properly. A television-quality picture uses a satellite transponder with a capacity of 6 million bits a second, but actually requires only half that or less. A single satellite transponder, in short, can carry one or two television channels, up to 1,000 two-way conversations, and more than 100,000 words of text a second. So, if people want to talk instead of sending teletype messages, they will only get about 1/50th as many words in the same channel capacity.[6] And if they want pictures, the cost goes up by a similar factor again.

As bandwidth gets cheaper, we can assume that people will begin to use it more liberally to obtain transmission of higher quality. Stereophonic radio broadcasting is becoming increasingly popular. People's expectations with regard to the quality of video transmission is also rising. When big-screen, true-color, full-stereo, full-width, high-definition TV begins to appear on the walls of people's homes, they will wonder how they ever put up with the television sets of today, just as we wonder how one could have put up with silent movies. Will they come to want 3-D? We don't know. And what will be the ultimate demand for videophone—with what quality and at what price? Again, we don't know.

The new broadband transmission media, efficient as they may be, are not costless. They will be more expensive than radio broadcasting in which a transmitter that costs a few hundred dollars using free bandwidth can be heard for hundreds of miles. But compared with the cost of other bandwidth today they seem likely to be much cheaper. The cost of paid-for bandwidth is clearly coming down. If optical fibers, satellites, and other such devices become widespread, and if we manage the spectrum wisely, there will be bandwidth at prices people can afford, for what from today's perspective may indeed look like a communication abundance.

Programming

In many fields, the biggest bottleneck to abundance of communication is not the hardware but the software. Consider education. There has been a great deal of loose comment about how education materials can be delivered electronically on demand to the home. However, an enormous amount of effort is involved in programming educational material for automated instruction. The time taken to write an article for publication as compared with delivering the same material from notes in a speech is about 10 to 1. But even writing it out in text is easy as compared with programming the same material for interactive automated instruction. Language is a slipshod mode of expression; to embody the same thoughts in a branched learning program requires a much higher order of precision. Programming is harder than writing by about 10 to 1. So preparing programmed educational material is perhaps two orders of magnitude harder than to stand up and talk before a class.

The same thing applies to entertainment. The cost of television is not the cost of transmitting it. It is the enormous cost of camera crews and actors working for days to prepare a few minutes of tape. The rule of thumb for cheap educational documentaries in the United States used to be $1,000 a minute. A commercial half-hour might cost $100,000 to produce. Yet by Hollywood standards those are grade B productions. Two hours of a Hollywood movie easily costs more than $15 million. If communication continues to grow at something like 10 percent per annum of output and maybe half that or less in consumption, it will not be entirely or even mostly by growth in programmed communication, and certainly not in programs intended for audiences of millions.

In the next chapter we shall turn to a likely change in the character and mix of communications from mass communications to group communications and to individualized communications. Mass communications are far cheaper, but as the cost of communicating comes down, some of the currently too expensive kinds of non-mass communications will become more affordable.

Chapter 5

Talking and Thinking among People and Machines

For most human beings, it is natural to talk and think at once. When alone, we often think in imaginary conversations. Mathematicians and artists may do otherwise, but most thinking involves trying out statements on imaginary audiences and adapting to their imaginary reactions.[1] When not alone, we communicate by a process of trial in which the flow of words starts before the thought is fully formed. One of the puzzles in linguistics is how it is that we form sentences, usually successfully, when we start them not knowing how we are going to end them. So interaction, and a sense of how the other person is responding, enters not only into communication but also into the very formation of the thought that is being communicated.

That may be how we converse, but most technologies of communication are one-way; they transmit only a preformed message. The written word, cinema, radio, and television have not lent themselves to dialogue. At best correspondence, talk shows, or citizens' band radio can be used for limited kinds of interaction. One can, of course, embody great thoughts in them. Books produced painfully and with much editing have become the vehicles for far greater intellectual content than casual conversations. But it is nonetheless the case that most media limit how most people are able to think in them.

Thinking while talking is the democratic mode in which most people are at home. Thinking without conversing is an artform of specialists. With one-way media there is a hierarchy. A specialist, a professional, a skilled person produces messages, while those who receive them are essentially passive. There are degrees of passivity; a student is far more active when she is reading and taking notes than when she slumps down with a beer before the television set, but to a degree in both situations she is just a receiver.

But many of the new inventions in communication technology may change all that. They often have been designed to overcome the inadequacy of one-way communication. Our era has witnessed the invention of "thinking machines" with which the human being interacts. Of course playing an electronic game on a television screen or sitting at an interactive computer terminal is not the same thing as engaging in a conversation. But newer technologies of communication are tending to fill in a gap between what were purely conversational modes (face-to-face and on the telephone) and what were purely one-way modes, like watching a television program.

The beginning of this book mentioned five trends in communication that appear to be changing society as much in this era as the printing press did five centuries before. The first trend, already discussed, is that distance is becoming no barrier to communication. Second, all kinds of communication are being increasingly interconnected in a single digital network. Third, the volume of communication is growing vastly, and channels, terminals, and programming for it are being provided at a cost that is not trivial. The technical fact behind the fourth trend, which is the subject of this chapter, is that because communication is becoming increasingly digital, the bits can be not only transmitted but manipulated by logical devices. Communicating and computing are merging; the logical operations that we normally call thought are being reintroduced into communication by devices—just as they have always been part of the process of conversation.

The Concept of Computer

It is misleading to talk about the computer as though it were a spatially defined machine standing alone, encompassed in a room. There are computers like that, many thousands of them; there was a time when even computer scientists thought of their devices in that way. But not any more. If the history of the computer has been written many times, what needs to be written is a history of the changing conceptions of the computer.[2]

The very name suggests an early misapprehension—that the computer is primarily a device for numerical calculations.[3] It was conceived in its early days to be a bigger, faster, more sophisticated electronic (instead of mechanical) calculating machine. In the 1950s the use of a computer for logical operations of a nonnumerical kind was something computer scientists understood to be an important area of research; mathematicians and logicians had understood the commonality of their domain all through this century. But the notion of using a computer in practical applications for editing, storing, manipulating, and retrieving text was only a glimmer in their eyes.[4] There were reasons for thinking about such applications. Early computer development in the war years had been wound up with work on cryptography. Then, after the mid-1950s, scholars found government sponsorship for work on the possibility of machine translation.[5] But outside of limited circles the image of the computer was still that of a super-fast desk calculator.

The modern devices that we call computers are much more than numerical calculators. What has evolved into a computer is a concatenation of four major elements, each with a history of its own.

Binary logic

One of these elements is the idea of expressing binary logic through on-off switches. In the pre-history of the computer there were indeed mechanical devices embodying the same logical concept. Automatic telephone switching systems were closely related; they are logic devices for selecting a destination by a combination of switches. But only with the work of Claude Shannon at Bell Labs, following Norbert Wiener's ideas, was a mathematical theory of information created—a theory which is at the foundation of modern computer science.[6]

In the decade of the 1960s, at the time computing was first impressed on the consciousness of the public, the logical operations were done in that part of the physical machine called the central processor; the notion that this was the heart of the computer was also conveyed by the term "main frame," which was applied to a piece of equipment in which, among other things, the central processor would be found. Terminals and auxiliary memory were attached to it. But all that was before the development of solid state logic with large- and then very large-scale integrated circuits (LSI and VLSI) on a tiny chip. Logic can now be located anywhere—in a hand-held calculator, in an input-output terminal, under the hood of an automobile to control the ignition, on a satellite in space, inside a television set to improve tuning or to make it a two-way device for games or pay-TV. Computing has become a distributed activity with pieces of logic put anywhere in the system, as desired.

Stored programs

The development of the stored program is what most historians would regard as "the invention" of the computer. Invention is a concept of dubious precision for any of these large systems. In the same sense, however, that we can refer to Bell's invention of a voice speaker and receiver as the invention of the large system that we now call the telephone, and in the same sense that we can call Morse's key and code the invention of the telegraph, we can call the development of the stored program the invention of the computer. The name associated with idea of the stored program is John von Neumann.

A thinking machine that operated only as the user punched each number and operation in turn (like an electromechanical telephone dial) would hardly have competed with the abacus. Stored in computers, by contrast, are programs that tell the computer what sequence of logical steps to go through to perform any given command that the user may give. Among the commands for which programs are stored are, of course, basic arithmetic commands like add, subtract, multiply, or divide. There are stored representations of each letter of the alphabet in some standard code, such as ASCII. The user can instruct the computer in English words, and the stored program translates the words and their letters to computer logic.

The stored programs mentioned so far are parts of the computer's own operating system. But making use of that operating system are also stored programs that do scientific or administrative operations that the user may have programmed for his own purposes. There may be a program for computing a coefficient of correlation, or determining the rate of return on an investment, or creating the index of a book. Such a program prescribes the operations step by step; for any given run of the program the user provides particular values for the variables on that particular occasion.

Storage medium

Where is the program stored? That brings us to the third element that makes up a computer, namely a storage medium. The program has to be stored, as does data to be put into the programs and results that make up the output. Any but the smallest of computers will have a variety of forms of storage. The important commands in the operating system that are used over and over again will be stored in a relatively costly medium such as an internal hard disk that permits extremely rapid access to them. Programs or data that are used only occasionally may be in a library stored on a much cheaper but slower device such as a tape. Data that belong to individual users will be kept on a cheap and detachable storage medium such as a floppy disk. All of these memories share the characteristic that locatable points in a grid can be set (like a switch) to the on or off position and thus represent in computer code the information that is being stored. Some storage media can be written on only once and are then permanently imprinted. They can be read but not rewritten again, and are called read-only memories. Other devices can be rewritten at will; those are called read-and-write memories.

Historically, the form of memory with which much early computing began was the Hollerith punch card. Before the computer came on the scene with its stored programs and electronic memory, the business and scientific tasks the computer later took over were largely done on what were then colloquially called IBM machines. The basic device was the Hollerith card sorter, introduced in 1890, which would drop cards into different pockets depending on which hole in a column of 12 positions had been punched. IBM moved from that business into the computer field, for it had perfected these electromechanical devices to a point where they did almost everything that early computers did electronically, and at first much more cheaply. They sorted on many columns at a time, counted while they sorted, did arithmetic, and responded to a program stored in a wiring board. So the transition to devices that used electronic components to do the same things was quite gradual. Indeed, even today in some business programs, the data are still being keypunched into Hollerith punch cards and entered into a computer in that way.

Punched paper tape, a heritage dating from stock tickers and teletype machines, was another early nonelectronic form of memory. It also is still used widely: for example, paper tapes are used to input a teletype message or to output it from a switching computer. Starting in the 1960s, however, these mechanical memory devices began to be by-passed. Keypunch operators began instead to type at a terminal that displayed the text on a cathode ray tube (CRT) and that coded what was typed directly onto some electronic memory, most often a magnetic tape cassette. The cost of such electronic memory has been steadily falling.

Input and output devices

The fourth element of a computer is its input and output devices, or terminals. The classic conception of a computer is illustrated once again by the way these are referred to; they are often called the "peripherals," as if there were a ring of them around the central processor. That is, indeed, an accurate description of the way that many computer systems are set up. But with time it is becoming less often accurate. As logic is being distributed throughout the system, and into the input and output terminals too, it often becomes hard to say what is the center of the system and what constitutes its periphery. The input–output devices are functionally at the edge of the system; they can take information into the system and pass information out of it. But the information they pass may go directly to or from other nodes of the same or contiguous electronic systems, and may be processed anywhere in the system.

Terminals, like memories, are of a bewildering variety. The most familiar are probably keyboard terminals, but they can be optical scanners, counters, manufacturing control devices, sensors like thermometers or barometers, clocks or cash registers. Whatever they are, they constitute an interface which translates between information that on one side of the interface is in computer-usable form and on the other side is in some form that is useful to a person or another machine.

As computer systems grow, and in turn get interconnected,

they become a diverse system. Logic, memory, stored programs, and terminals are scattered throughout. The system becomes part of the largest machine that man has ever constructed—the global telecommunications network. The full map of it no one knows; it changes every day.

Computers and Communication

We have described this fourth of the great technological trends of our times not so much as an increase in computer usage (which of course it also is) but as an increase in the interactive capability of the communications system. Logical reaction becomes part of its capability. The machines react with intelligence, and so give their users some of the interaction that was previously available only in conversation.

What has made it possible for a physical network of equipment to interact with its human user is the marriage of telecommunication with computer logic. This coupling is achieved in part by attaching what could also be stand-alone computers to the network and in part by incorporating digital logic into the telecommunications system itself.

For some time the telephone switching exchanges have been going digital. As with any major change in the technology of the massive telephone system, this has been a long, slow process. Virtually every country is moving to an all-electronic digital switching system. In the United States, the first electronic exchanges, AT&T model ESS-1, were installed in the 1960s. The first advanced fully electronic exchange was initiated in Chicago in 1976.

Digital exchanges are more reliable, cheaper, better adapted to data transmission, and much more flexible. Being a large computer, the electronic exchange operates under the control of a stored program. That software can be changed without changing the hardware at all. It can be written to do many complex and potentially useful things, which in a nonelectronic switching system would have to be expensively wired into the hardware. For example, where electronic exchanges have been installed, customers can obtain call-forward service, so that if the customer goes away from one number to another location, incoming calls are automatically transferred to the other number.

A second factor in the marriage of telecommunications with computation is the use of "distributed logic." In the 1960s it was not clear that the wave of the future lay in this direction. At that time people were impressed by the economies of scale to be found in computing. It was observed that the power of a computer, as measured by the number of operations it could perform per second, seemed to rise as the square of the cost of the computer. A computer that cost twice as much could do four times as much work as its half-size brother. One big computer was more efficient than many small computers. What followed was the forecast that there would ultimately be large information utilities, like the telephone or electrical utilities, to which everyone would connect to do their computing. That notion aroused both enthusiasm and fear. If there were to be grand computer centers serving thousands or millions of customers, the challenge to design an efficient, reliable, and secure system was enormous. It could not be allowed to fail any more than the electricity or phone systems can be allowed to fail. Its efficiency would be the key to the productivity of society. It had to be flexible to meet an enormous diversity of needs, and it had to protect the privacy of each user's files. The fears that were engendered by this image of the great master computer were mostly centered on the notion of a gigantic national data base in which all records on everyone would be kept. Indeed, in the late 1960s there was considerable national agitation in the United States, Sweden, Canada, and elsewhere about the dangers to privacy inherent in the great data archives that were assumed to be coming.

By the 1970s the image of the future had changed. The miracle of miniaturization and large-scale integration led to the marketing of minicomputers that could do the same things which only a giant, multimillion dollar computer could do a few years before. The notion of the computer information utility was challenged by a new doctrine, one which turned out also to be only half correct. The argument went more or less as follows: In the first place, the economies of scale in computing, while still present, were far less important in the age of very large-scale integrated circuits than they had been earlier. With the equivalent of a large computer on a tiny chip, a minicomputer was not as drastically inefficient compared with a large computer as it had been. Second, and even more important, whatever efficiencies a large computer would have in computing would be swamped by the extra costs of communications in order to do remote computing on the large machine. The costs of computing were falling far more rapidly than the costs of communication. While in the past computing costs dominated communications costs in a user's bill, in the future it would be the communications portion of the total bill that would dominate, and so it would be the communications costs that the user would wish to minimize. Thus the argument was that in the race of falling costs between communications for remote computing and investment in minis, the minis were clearly winning. The picture thus predicted was of a world in which there would be a very large number of localized minis, and as a result data communications would decline in relative terms.

This prediction was correct in seeing an enormous growth of minis (and later micros) on the horizon and the extensive displacement of large computers by them. But it was wrong in forecasting that this growth would reduce the amount of data communications. Today, minis and micros are chattering with each other even more than terminals would have communicated with a great computer utility.

The prediction that the growth of minis would reduce communication was incorrect because it looked on computers as calculating machines. It focused too exclusively on the internal economies of computation, on how much a particular set of calculations cost, adding together both computing bills and communication bills to do it. It did not take account of what people would be using their computers for and the total costs involved in those activities. Specifically, it did not consider that a large part of what people use their computers for is communication.

For example, when the ARPANET (Advanced Research Projects Agency Network) was developed, the expectation was that people would use it to take advantage of especially good software that might be running on a computer elsewhere; people were expected to use it to do computations that they could not do at home.[7] There is little use of that kind because once users learn their own programs and machines, they rarely find it worthwhile to take the time to become familiar with another set that has its own special idiosyncracies. But the ARPANET has been used a great deal. It has been used for communication. It has created a community of scholars who work together and exchange experiences and information.

So too with minicomputers. One of the strongest reasons for their growth has been to improve communication. Some of them are explicit message-switching computers. More of them do some processing locally but also serve to transmit the results to a center. For example, when Avis set out to design its Wizard system for computerizing its car rental agencies, the initial notion was that all the terminals could work by communication lines to a central computer. It turned out that the costs would have been prohibitive. However, by making each of the terminals an intelligent one with a minicomputer, most of the calculations are done on the spot, and only the essential information is transmitted to the central computer. Such a system increased computer communication, despite the distributed computation at the terminals.

The main reasons for networking are rarely to save money on computing; they are usually to save on the much greater human costs that are exogenous to the computing process itself. The costs of preparing data, checking it, editing it, maintaining security, and so on are apt to be far greater than the costs of data processing as such. Stockbrokers may find it worthwhile to get mini- or microcomputers rather than using only a time-sharing stock quotation service, but they will still get the stock quotation service for current quotations. The computer will work on that data.

In many situations the choice is not either to distribute computing to many locations *or* to communicate with a central computer; the choice is to both distribute computing *and* to communicate. For each particular operation a cost minimization calculation can be made as to whether to do it locally or to communicate the data to a larger machine for doing the calculation, but for the operation as a whole, the very fact that many calculations will be decentralized to local equipment means that it is all the more important to keep the various machines in touch with each other.

Interactive and Individualized Communication

In one way or another, the programmed logic that can be built into modern electronic communication is reducing in part the passive uniformity of mass communication. Just as computer-controlled assembly lines can vary the product in a way that would be prohibitively expensive otherwise, so too computer-controlled media production can bring into the realm of economic feasibility kinds of communication that take some account of the individuals to whom they are addressed.

The mass media revolution of the nineteenth and twentieth centuries did for information what the mass production revolution did for physical goods. In both instances it was found that machine production of a uniform commodity under factory conditions could be done very cheaply. The rotary printing press produced penny newspapers by the hundreds of thousands. The economics of broadcasting likewise depended upon reaching millions of people with a uniform message. No one could afford to run a broadcasting station to meet the special interests of a few hundred listeners. The telephone was the only major communication development that allowed for individualized communication, and that fact neatly separated the business of the phone company from that of mass communication.

Today, however, new technology is allowing the editors of mass media to tailor their contents to the interest of small specialized audiences. For example, newspapers and magazines will sell ads to appear in just that part of their circulation which is distributed in a certain district. Such variations are possible because of the flexibility that computer-controlled composition allows.

Editors are not the only ones who can use computer logic to control communication. Among these are computerized information retrieval systems, such as the stock market systems and the airline reservation systems. Inventory record systems have become common. Much smaller, but also growing rapidly, are the scholarly and general information systems such as Chemline, Nexis, and various bibliographic search systems. As such query-answering services grow, they will begin to replace mass media in part. Instead of reading a journal that is edited to meet the needs of all its readers as best it can, researchers may retrieve from the mass of research reports only that subset which matches their particular interest.

It would be unrealistic to expect that passivity of personality will somehow vanish under the wand of new technologies. The most passive media will remain among the most popular. But it may make some difference if future generations grow up with devices at hand that allow them, at reasonable effort, to actively involve themselves in meaningful endeavors rather than simply accepting a mass product. The new telecommunications technology will at least permit people to talk and think not only with each other but also with media machines.

PART II

SATELLITES, COMPUTERS, AND GLOBAL RELATIONS

Chapter 6

Communities without Boundaries

As the cost of communicating becomes independent of distance, the question arises, What are the consequences of a technology without boundaries? How have governments reacted to multinational communication, and what can social science tell us about its impact?

Historically, human activities have been structured by contiguity. The basic social, political, and economic units were villages or towns, aggregated into provinces, nations, and regions. Telecommunications and air transport are changing that. New York and Los Angeles interact more with each other than with the Great Plains in between. In various professions, specialists in "invisible colleges" write, call, fax, and exchange reprints with one another around the world.

With a satellite, the communications distance between all points within its beam has become essentially equal. In the non-Euclidean communication plane that results, no point lies between two other points. Boundaries partition nothing. The topology of commerce, government, and social life may change to reflect that space warp. As communication between countries becomes as technically easy as communication within them, the division of labor and interdependency among nations is affected. Already the news services that everybody uses, the movies everybody watches, the information bases everybody accesses operate increasingly on a global basis. Multinational organizations locate units wherever there is a comparative advantage for an activity. Some countries will use the new technologies of communication for international leadership. Others will find business activities moving abroad and their information and entertainment production slipping away from them.

No government happily accepts a loss of control over what happens within its boundaries. The normal response to foreign influences is to build walls. When Italian shoes or Korean tankers capture a market, the instinctive impulse is to shut them out. Yet for two hundred years free-trade economists have argued that excluding foreign goods is self-destructive. Under most circumstances, economic advantage goes to the country that opens its arms to the world market, and decline comes to the country that tries autarchy.

How far does this argument apply to the flow of communication? As is the case for consumer goods, growth of international communication has stirred protectionism. In this chapter we will use the social sciences to shed light on the question whether those who slam down the lid on communication of foreign ideas will gain or lose thereby.

There are no isolated civilizations in the world, nor have there ever been. Ralph Linton wrote, in 1936, that no culture "owes more than 10 percent of its total elements" to its own inventions.[1] The rest has gained through diffusion from other cultures. Sometimes the very things that symbolize a culture's identity turn out to have resulted from foreign intrusion. For example, in Machiko in Japan and at Atitlan in Mexico particularly beautiful folk pottery is made, yet in each case a foreigner initiated its further development to help create a market for cottage crafts: for Machiko it was the British potter Bernard Leach in 1904; for Atitlan it was a Franciscan priest in the nineteenth century.

Though there are no truly indigenous cultures, in the past the accretion of foreign elements was clocked in generations or centuries; today it is years or even microseconds. The world is getting smaller, and its implosion is exponential. Steamships, airplanes, telegraph, radio, and telephone accelerate all sorts of relations: trade, alliances, colonization, tourism, cultural exchange, and diplomacy. These interactions approach being instantaneous regardless of national boundaries.

Today, anyone can talk to anyone within a single city; soon one will be able to communicate with anyone throughout the world. But who will actually communicate with whom at that time? Technology will pose no bars, but social customs and structures most certainly will limit the growth of "community without contiguity."[2]

The Spatial Reorganization of Activity

Communications technologies are beginning to have visible effects on the spatial organization of human activities both within and among nations. Before instantaneous communication to a distance was made practical by the telegraph and the telephone, senior diplomats were called "ministers plenipotentiary." The name is significant. In those days when an exchange of letters with home could take months, the ruler who sent an ambassador off had to bestow full power on him to negotiate, bargain, compromise, and commit his country without checking back.[3] Today, an ambassador is no farther away than the other end of the telephone line, and he does not change a comma in an agreement without first checking.

Ship captains used to sail off, perhaps for two years, with a substantial sum of money from the owner and an extraordinarily general letter of authority to buy and sell as the captain thought best. Today, satellites provide instantaneous telex, voice, and video communication to those tankers and other high-value ships that have been equipped with satellite antennas. That is the first step toward a system in which ships could be navigated under remote controllers, as airplanes are in the commercial traffic lanes today. An earth observation satellite linked to computers can measure the waves and currents in the ocean below it and calculate courses and optimal arrival times, while a captain can only observe conditions immediately around him.[4] So the days of the free-roaming captain with authority over all he can see may be numbered.

When long-distance telephones first came into use, they immediately changed the structure of management. "The Foreman of a Pittsburgh coal company," wrote Herbert Casson in 1911, "may now stand in his subterranean office and talk to the president of the Steel Trust, who sits on the twenty-first floor of a New York skyscraper."[5] Today the example could be global. In many corporations private-line networks are linked to management information systems, which process daily reports from branches all over the world into the company's computer.

From such facts the plausible but fallacious conclusion is easily drawn that telecommunications and the computer have increased centralization. That is at best a half truth. Let us review what actually happened in the case of the telephone.

Centralization, decentralization, and the telephone

The long-distance telephone had several related effects: undercutting local managers while at the same time enabling companies to grow and spread and loosen hierarchic relations.

The phone reduced the authority of corporate officials, bankers, and politicians in the field. Branch managers used to be partners in the bank and made funding decisions on their own; now they are employees.[6] On the subject of political campaigns, *Telephony*, in 1906 (p. 364), reported that the telephone "has curtailed the functions and responsibilities of a district manager as the cable has those of an ambassador."

At the same time, the telephone helped organizations to grow in size and spread. It used to be that most companies were in a single main business and centered in one main city. Sears Roebuck was a Chicago company, Fiat was in Turin, and Krupp in Essen. Indeed, each had plants and offices elsewhere, but in a star-like network with control in the center. Today, conglomerates are not like that. Their centers may not even be in one country. Their nationality is increasingly diffused in dispersed management, ownership, and production.

The dispersion of conglomerates began with the separation of corporate offices from industrial plants at the turn of the century. In the mid-nineteenth century most large industrial companies had a major plant in a single location, usually near water. Small plants and distributors, however, were concentrated in single-activity neighborhoods in the middle of each city; so if one wanted to do business with, say, furniture makers, one went to the furniture district. The makers, sellers, and material suppliers were all there within a few blocks; one could walk up and down to find what one wanted. With the coming of the telephone, all this changed. Corporate offices moved away from the factory, which could be adequately controlled by a phone call to the hired manager; the president moved downtown, where he could have face-to-face meetings with bankers, suppliers, and customers. Traders could move away from the high-rent, single-product neighborhoods, either out to where their customers were or into offices in the new skyscrapers in the heart of downtown. And so downtown changed from a collection of specialized neighborhoods to a dense concentration of business offices engaged in commerce with one another. A group of bankers and entrepreneurs interacting with one another in lower Manhattan could put together deals for investment and trade and could still keep control over their different enterprises in scattered locations by calls to the local managers. At its peak, in the years just after World War II, 139 of America's 500 largest corporations had their headquarters in New York; in France, the United Kingdom, and Japan, the concentration in the central city was even greater.

But what telecommunications spawned at one stage it undermined at another. For years now corporate executives have been debating whether in this day of computer networks and facsimile machines they need to pay the high rents in downtown Manhattan, Tokyo, or London, force themselves and their employees to commute an hour each way to get to work, compete for the high-priced labor force of the metropolis, and put up with the frustrations of city life when they could conduct their business just as efficiently from a less congested rural setting. Government managers, too, have been asking whether their booming activities have to keep moving people into the capitals, or whether many offices could not function just as well in dispersed locations.[7]

Another effect of the telephone was also to loosen hierarchical layering. As Marshall McLuhan put it, "The pyramidal structure . . . cannot withstand the speed of the telephone to bypass all hierarchical arrangements . . . Today the junior executive can get on a first name basis with seniors in different parts of the country."[8] The telephone allows people at different levels to talk to one another informally, bypassing their intermediaries. It permits people in different branches to confer without necessarily following the chain of command.

This informality infuriates those who get bypassed. When President John F. Kennedy phoned his friends in government departments, his cabinet secretaries fumed. As things are now, a middle-level embassy officer abroad has various ways of communicating with officials he knows in Washington. He can write a formal airgram, which gets vetted, nominally by the ambassador whose name appears on it as the official source, though probably only by the Deputy Chief of Mission. It gets mailed in the pouch and may take a couple of weeks to reach its final addressee. If time is important, the same thing can be done by cable; the text is written out, and after vetting it gets delivered to the teletype room to be keyed in by a telegrapher; that message should take only a day at most. A personal letter to a bureaucrat at home who happens to be a friend could bypass the official system altogether. The telephone is also available, but that is expensive and is not encouraged unless important; because of cost, the government leases insufficient circuits to handle heavy traffic.

It has often been urged that terminals on embassy staff desks would enable a typed message to appear immediately in the terminal of any similar official in Washington. Such a message facility would have the flexibility and instantaneity of the telephone but be cheap enough to be used freely. It would bring the embassy staffs into the same situation as their fellow bureaucrats back home in the United States, who have telephones at hand with which they can talk to anyone they wish, without concern about the cost or availability of lines. Yet this suggestion meets fierce resistance on the grounds that it will prevent effective control and coordination by the ambassador of his team. Anyone in the field would be free to reach anyone in Washington. Back home, this situation does not appear to get out of hand; reasonable norms control whom people actually call directly. But in the foreign services, telecommunication is feared as an instrument of decentralization that reduces hierarchic control.

In discussing the social impact of telecommunications technology, it would be better not to talk of centralization or decentralization at all. It is tyranny of language to insist that a process with several different aspects has to be characterized as fitting or not fitting a particular word. We can describe what communications facilities do: they make for larger organizations, but with more dispersed and varied activities, with less autonomy on the part of field agents, but also with less rigid hierarchy. Whatever we call it, the flow of interaction both nationally and internationally resembles less and less a set of spokes with a single hub. New means of communication permit all sorts of multilateral flows.

The multinational flow of communication is present not only in business and defense but in science and learning, too. Facilities for the collegial exchange of information (universities, academies, libraries, journals, conferences, and information retrieval systems) are the institutional base for science, and all of these operate in a highly cosmopolitan way. Note, for example, the growth of international congresses. Two or three congresses or conferences took place annually in the world from 1838 to 1860. Beginning in 1900, a hundred or so congresses were held annually, and in 1910 two hundred. By the 1970s, the number of congresses, conferences, and symposia organized or sponsored by international organizations exceeded 3,000, not including all the purely administrative and working group meetings.[9] There are more than 100,000 scholarly and technical journals in the world.

Scholars have always been part of a free-floating community whose ideas transgress national frontiers. This becomes even more true as electronic communication makes the functioning of invisible colleges and visible meetings easier.

The Global Flow of Mass Media

Charlie Chaplin, Big Bird, and Marcel Proust may have little else in common, but they are all figures in a cosmopolitan culture that, for the first time in history, embraces the globe. From the turn of the century until the present decade most of the growth in international communications flow was in the mass media. In this growth, costs have been decisive. In the Middle Ages, manuscripts were so scarce and so valuable that scholars would wander through Europe from monastery to monastery to read books in the places where a copy happened to be kept. The cost of diffusion of knowledge was the cost for a person to travel. With the coming of printing, the scholar could stay put and have the books delivered. Until the telegraph, physical transport of paper remained the primary method of diffusion of knowledge. The telegraph, while it added an option, was far too expensive to be used as a means of expression. No one distributed essays via telegraph. It was not extended to people's homes but was used as a business machine. It was for orders, payments, travel arrangements, and brief personal messages, but generally not for lengthier self-expression.

The telegraph was also used for news, but only on a wholesale basis. Wire services sent bulletins to newspapers, which reprinted them for widespread distribution. The telegraph made mass media more cosmopolitan, but economics still dictated that the message delivered to the public be printed on paper and be limited thereby in the radius to which it could be economically sent.

The twentieth century added new means of distribution that

freed voice and drama from their local roots: cinema, the phonograph, and radio. These could be delivered around the world, but in their early forms they remained too expensive to deliver to the public except as mass media, first in places of public assembly and then in homes. After 1896, movie houses were built where people could assemble to see a film. Starting about 1897, phonographs and records were manufactured cheaply enough to be used at home rather than in a nickelodeon. In 1920 radio broadcasting emerged, and after World War II the motion picture moved into homes via TV sets which in a rich land virtually everyone could afford.

Movies were at first mostly made in New York and then in Hollywood and then shipped around the world. Phonograph records were produced in Europe as well as in America and exported, too. Radio broadcasting, on the other hand, was limited in its initial reach to the scope of domestic networks. Foreign radio became important with short-wave broadcasting in the 1930s. Television from its start exported and imported many movies, and then with the development of videotape recorders around 1957, TV became, even more than before, a major international rediffuser of culture.

Except for short-wave radio, the main way before the 1970s in which communications media could be transmitted cheaply enough for international mass distribution was as single copies to wholesalers, who then put them into a domestic distribution stream. For half a century or more it was inexpensive enough to put a movie or videotape in a can and airmail it abroad; only now do we see the prospect of transmission facilities becoming available whereby the individual citizen in the audience can be reached from anywhere directly, either by a mass medium or by another citizen.

Let us trace the sequence of internationalization and how it affects each medium differently, according to the technology of its production.

News services

The first international news service was established in 1848 by Paul Reuter. He had been earning a living in Paris translating articles and commercial news and sending them to papers in Germany. There was a gap between the new German telegraph line at

Aachen and the French and Belgian line at Brussels which Reuter filled with carrier pigeons. As that gap was filled and telegraph lines spread, he adopted the slogan "Follow the cable/telegraph." He moved to London in 1851, where Reuters has been based ever since. Occasionally, for major events Reuters would incur the expense of a journalistic spectacular by telegraphing a lengthy text in its entirety, the first time for a major speech by Napoleon III in 1859; but the bulk of wire service material in the nineteenth century consisted of bulletins. Newspapers would run them in a special box headed something like "late telegraph bulletins." These items, often in bold type, informed the reader of any crisis of the day, but extensive analysis had to await the mails. Half of Reuters' business then, as today, consisted of high-value but condensed commercial information like price quotations.[10]

In both England and the United States, the telegraph companies themselves tried to go into the news-service business. Western Union, the telegraph giant, saw in its wire network the natural base for the business. Its telegraph agents everywhere could be news collectors to feed its transmission facilities. Associated Press's news sources, on the other hand, were better than Western Union's. Its suppliers were its member newspapers, whose reporters had advantages over Western Union telegraphers.

The practice among newspapers, in disregard of copyright, was to lift stories from one another. In the days before the wire services, this was the way newspapers got their news. Newspaper publishers would exchange free copies of their papers with each other. To facilitate this exchange the early postal laws—for example, a law of 1792—provided for free mailing of newspapers to other newspapers. With the coming of the telegraph this franking privilege lost its significance and was abolished.

In many countries what developed instead were concessionary telegraph rates. Where telegraph services were state-owned, the reductions were often massive and politically based. Governments were willing to suffer a loss to keep the political good will of the press and to gain some leverage over it. And commercial telegraph and cable companies provided discounts, because the press was such a large customer.

Special telegraph rates for the press have become obsolete today. By now very little newspaper traffic is sent by telegram. The great news agencies have global networks of private lines. Even traffic from isolated locations off these networks, like a reporter's story from the field to the nearest bureau, is more likely to be telephoned or telexed than telegraphed. Until a few years ago most of the private lines were of teletype speed and the terminals were typewriters. That has changed; on busy routes news agencies lease and multiplex high-speed circuits and transmit the text into computer storage for editing and retransmission. Yet as late as 1973 the press had leases on 90 teletype circuits between the United States and overseas, but only 6 voice-grade lines.[11]

There are now scores of news services in the world, most of them national, often government-owned, but the great bulk of the traffic is handled by five agencies: Associated Press (AP), United Press International (UPI), Reuters, Agence France Press (AFP), and Tass. The great agencies have correspondents scattered over the world and news-exchange arrangements with the numerous weaker national news agencies, some of which also have exchange arrangements among themselves.

The great news services, each originally set up to serve the needs of a particular national set of newspapers, have in the last couple of decades increasingly evolved into multinational organizations. A large part of their revenue comes from serving the world press.[12] Their new computer-managed facilities allow much more diversification of service to different areas; the menu of different types of services to which a given paper can subscribe from any given agency is growing. The agencies compete vigorously for foreign clients and so to some extent must become sensitive to what their customers want.

Yet those countries that are not headquarters for one of the great multinational news services are dissatisfied. Many claim that news is reported with a bias alien to their needs and values. Most often the argument is made that the major news services are in developed countries and fail to meet the needs of the developing lands. The charge is that the flow of news is unbalanced: more news is sent about Europe and America to Asia and Africa than about Asia and Africa to Europe and America. The charge also is that the news being sent covers neither problems of development nor non-Western cultures and further that ideological and commercial propaganda gets into the news. While there may be less truth to some of those assertions than is claimed, they are not all groundless and they have had much impact. (We will return to this issue again.)

One new trend in news distribution must be noted: With the falling cost of electronic transmission, the press services that started out to deliver news as wholesalers can increasingly bring news by direct electronic transmission to the final customer. For a long time private organizations, as well as media, have been allowed to subscribe to wire service tickers. In fashionable clubs, luxury hotels, brokers' offices, the White House, or large corporate headquarters, one could see the teletype ticking away with the AP wire, or Reuters, or AFP. Long strips of paper were tacked to the wall. This was not seen as significant competition by the newspapers that sponsor the press services, and it was an added source of revenue to the agencies.

But now, with high-speed lines and computer technology, all sorts of new options open up. Some of the large press agencies have developed commercial services for individual customers, following Dow Jones' 1967 joining with AP to form the AP-Dow Jones Economic Report.

Equally significant is another possibility inherent in the business services. With the development of direct transmission of information to subscribers, the services can evolve from simple information services into *transaction* services. In the United States, on-line stock quotation services are evolving into computerized stock exchanges that could threaten, in the long term, to eliminate the formal exchanges. Needless to say, the possibility is causing alarm in the City of London and on Wall Street.

Books and magazines

Even while press services and syndicates made information available from afar for printing locally, the physical export of books and magazines has also been growing. Before World War II there was but little international circulation of magazines aside from scholarly and technical journals. Sea mail was too slow to be useful for topical publication. After World War II, for about fifteen years there was a marked discrepancy between the United States and Europe in the availability of slick paper, modern printing plants, expensive journalism, and the resulting glossy popular magazines.[13] *Time, Newsweek, Life, Look,* and the *Reader's Digest* found lucrative markets abroad. Shipping costs and delays and the problem of language led the successful magazines in that export market to print local editions overseas.

However, by the 1960s and 1970s magazines were appearing in numerous countries and language regions. Publishing empires such as Axel Springer and Bertelsmann in Germany or Asahi in Japan have far outdistanced the growth of American-based magazines in their countries. While magazines may copy the formulae used by successful ones abroad, editing and production are increasingly domestic or regional. Globalization may be observed in the transmission of electronic information, but any medium that requires physical transport of heavy paper is likely to use local or regional production and distribution; and that gives an edge to independent entrepreneurs in those locations.

Phonograph records and tape recordings

Like books and magazines, phonograph records are heavy and expensive to ship, so plants to stamp them are widely dispersed. However, the concentration on top pieces and artists in records is greater than that in print, and so the market is dominated by fewer firms. Their material is multinational. The shelves in any record store present artists from all over the world; the original producing companies, too, are diverse in origin. Licensing, however, allows the major distributors to extend the market.

At least that was the way before the tape recorder. Now the capital plant needed to launch a career is minimal. Any performer can make a master tape, and anyone who listens can make a tape for himself. "Record stores" in Hong Kong have records there for customers to play; but when a customer makes his choice he does not take the record with him; he hands it to the clerk to copy onto a tape cassette, copyright be damned. Whatever international concentration the domestic record firms were once able to impose on the world market, they can do so no more in the face of widespread piracy.

Cinema

No medium has a more multinational history than the motion picture. Thomas Edison invented a kinetoscope in 1894.[14] It was a device one peered into—a kind of nickelodeon such as one finds today in porn shops or amusement parks. In 1895 a Paris group developed the idea of a movie projector. Edison had so little faith in his device that he did not spend the additional $150 for an international patent, so the French competition with projectors compelled him to move ahead with his own vitascope projector.

Regular production of dramas on film and their distribution in theaters began with *The Great Train Robbery* in 1903. From the beginning, distribution was international. The pictures were, after all, silent. The dominance of Hollywood in the first third of the century epitomizes one pattern of international activity. Rarely, if ever, has one country exported its culture as massively as in the heyday of the Hollywood movie. Westerns, gangster films, jazz, and variety shows all became part of world folklore. United Artists, Warner Brothers, RKO, MGM, and Universal came to depend critically on export markets for their margin of profit. By the 1960s, according to Thomas Guback, foreign earnings were 53 percent in film rentals.[15]

But as a result of Hollywood's dependence on its foreign markets, its empire was a passing phase. After the 1930s, Hollywood lost its franchise on the world cinema market. Its very success bore the seeds of competition. Hollywood pictures were made for the American audience, but they had a universal human appeal. As a result, local entrepreneurs had an incentive to set up movie houses everywhere—luxurious palaces in rich cities or benches in a shed in developing villages. A movie-going public was created that acquired the habit and paid for the products of the industry.

In Europe first, and then in China, India, Japan, Egypt, and Mexico, local talent discovered that they could make movies, too. In poor countries they were not paid like the Hollywood stars or union cameramen, and so production, being labor intensive, turned out to be cheap, if not slick. What is more, the local audience liked the local product; it was attuned to their taste. So the original inflow of Hollywood films, by opening movie houses, laid the foundations for a local industry. Far from Hollywood's activities being self-perpetuating, they turned out to be self-eroding. Today American moviemakers often depend upon co-productions with foreign makers to succeed.

The technology is increasingly favorable to dispersion of production. Professional-quality work can now be done with low-cost cameras and editing equipment. With $3,000 and some talent, young filmmakers can now go into production. Some of the most interesting work being done in Japan, the United States, and

Europe is by underground artists. Of course the day of the mammoth production studio with millions of dollars' worth of equipment, a crew of dozens, and production costs of thousands of dollars a minute is not over; this system produces spectaculars that cannot be made in any other way. But the day of the major studios' oligopoly over serious production has passed.

Radio

Before broadcasting emerged, radio telegraphy was thought of as essentially an international medium. Still earlier, prognosticators had foreseen such uses for radio as enabling lighthouse keepers offshore to communicate with the mainland. But once Marconi showed the reach of long waves, the value of radio for international communication became obvious.

For broadcasting, the merit of long waves—their following the curvature of the earth—was a liability. Long-wave stations from great distances interfere with each other. Medium waves had the asset that stations in different countries or language areas could be on the same frequencies and not interfere if they were a few hundred miles apart. The basic pattern of broadcasting that emerged in most countries was a strong station or two beaming from the national capital, supplemented in medium-sized countries by a few weaker regional stations. The United States was large enough and federalist enough in philosophy to follow a different pattern. The Radio Act of 1927 and the Communications Act of 1934 made localism the basic structure of American broadcasting. Each fair-sized town was assigned one or more frequencies for stations; they were to serve their communities.

Still shorter waves, it was found by the 1930s, travel in straight lines like light, providing a smaller radius of possible reception. However, as radio amateurs found out when they were shunted to those frequencies thought to be useless, shorter waves also bounce off the ionosphere and so can be used for extremely long-distance transmissions. Thus, in the 1930s short-wave international broadcasting began. Born in a decade of ideological conflict, short-wave broadcasts from the start were used for psychological warfare. By World War II every major power was sending its message abroad.

Successful examples were the broadcasting of Voice of America,

Radio Liberty, and Radio Free Europe to the Soviet Union and Eastern Europe.[16] The evolution and persistence of the dissident movement in those countries was aided by the information these foreign broadcasts brought and their reassurance about world attention to their movement.[17] Intellectuals learned about current world debates, and casual listeners heard a new version of the news. In the 1960s, about one sixth of the Soviet population listened to foreign stations on an average day.[18] In contrast, only about 2 percent of the American public tunes in to short-wave broadcasts at all.[19] Few families in the United States even have short-wave sets, in contrast to one half of all Soviet families. Rightly or wrongly, Americans think they are getting the essential facts via their own media. People do not turn to foreign broadcasts when they are satisfied with what they get through normal channels. Listening to illicit political broadcasts becomes important only in a milieu of repression.

Americans, who use short-wave radio so little, can easily underrate its significance in those places where the domestic media fall short in the eyes of the people. Short-wave's past powerful attraction in Eastern Europe shows what impact a direct international flow can have when it meets a real need.

Citizen band radio's success in the United States reinforces that lesson. In contrast to short-wave broadcasting, which has historically been one-way preachment, CB's success shows a felt need in American life for a two-way forum. The CB audience consists of people who are gathered together, although separated in space, in a mutual desire for social interaction. Telephone is also two-way but usually requires a targeting of particular persons called; to call them is a proposal that they chat on your terms. CB is to telephone as a cocktail party is to an interview.

Today CB licenses are deliberately restricted in range and are available in only a few countries. But consider what may happen if, at some time in the future, the CB model is put together with that of short-wave radio. Suppose CB existed with a global reach; new kinds of international communities would form.

The model already exists to a small degree. A worldwide CB community would be an extension of the existing community of radio amateurs (hams) who have talked and morse-coded to each other since even before the birth of broadcasting. But under current rules, hams have to pass often rigorous tests of technical and operational proficiency, their equipment is relatively expensive, and the subjects on which they may converse are limited.

If technology, spectrum allocation, and government rules would permit short-wave remote conversations at the cost and ease of CB, the international results could be as remarkable as they have been with national CB. Indeed, once several million Americans learned to use CB and realized its technical limitations, many of them sought to upgrade their range (often illegally), and many applied for amateur licenses.

Television

Television had been the target of strident accusations of cultural imperialism. It is blamed for introducing commercialism and consumerism, violence, and pornography. U.S. television, in particular, is blamed for spreading those evils to countries that held different values. Most of the world is still at a stage of media development in which it imports much of its programming from a relatively few production centers, most notably the United States. An early documentation of this fact was the 1973 University of Tampere's *International Inventory of Television Programme Structure and the Flow of TV Programmes between Nations*. The study estimated total foreign sales of U.S. program material at 100,000 to 200,000 hours, of which about one third went to Latin America, another one third to the Far East and East Asia, and the rest mainly to Western Europe. The proportions in dollars were different, because the rates charged in Europe were much higher than those for less developed countries.

In general, the dominant role of American material is declining, as one would expect. The United States was the first country to go into massive TV production, but gradually other countries built up their own capabilities. Today, there is enormous variation between countries in the proportion of programs imported. A few countries show extraordinarily low import figures in the 1973 Tampere study: United States, 1–2 percent; USSR, 5 percent; Japan (NHK only), 2½ percent. But about half of Latin American programs were of foreign origin; a similar figure existed in the Middle East and in Asia, excluding Japan and China; in Western Europe, it was 30 percent, and in Eastern Europe, 24 percent.

The foreign television shown in those countries consists of programs which their own broadcasting authorities have chosen to buy. The foreign programs are mostly shipped physically for redistribution by a domestic TV station. Satellite distribution of feature programs has lagged far behind the transborder distribution of current events and sports. Spectrum allocations do not favor transborder television. By the time television came on the scene and needed spectrum allocations, the long- and medium-wave portions of the spectrum were already occupied, leaving the Very High Frequencies (VHF) and Ultra High Frequencies (UHF). At such high frequencies transmissions are basically line-of-sight. They do not reach much beyond the horizon. So the effective radius for TV transmitters is below 100 miles. Therefore, there was little international broadcast TV in the past, and whatever foreign programs were shown were those chosen by domestic stations as appropriate to their requirements.[20]

Will that change? Will technology permit direct international reception in the future instead of redistribution through domestic transmitters? Such a change could come about in a variety of ways. The first and least likely would be by a reallocation of spectrum to TV at frequencies that permit long-distance transmission. Given the fierce possessiveness of the present frequency assignees, the amount of spectrum that color television requires, and the political opposition to direct international TV broadcasting, we can dismiss that as something that is not going to happen outside of collaborative environments such as Europe. A second way that direct international TV could come about would be by direct satellite transmission to home receivers. Yet few communication issues have raised more international controversy. Chapter 8 deals with the matter of direct satellite broadcasts at some length. Suffice it for the moment to anticipate the conclusion that direct satellite transmission is less likely to be as important a means of direct TV distribution as a third way, namely, transborder satellite distribution for redistribution by local broadcasting stations and cable TV systems.

Cable television, cassettes, and video disks

Cable television might seem to be a purely domestic medium and to have no international implications. But this is incorrect. Cable systems can be networked, both domestically and internationally, and in terms of both hardware and software. Such networking of cable systems is in full swing in the United States and Europe.

Subscribers to cable have many options (systems with more than 50 channels are common) and find far more variety available to them than was available by over-the-air reception. There is, of course, much duplication and appeals to the broad center of program preferences. Nevertheless, the expanded diversity at any time of the day has been extraordinary.[21] Video cassettes or video disks can also give specialized and minority audiences, poorly served by over-the-air TV, the special fare in which they are interested. By one means or another, then, it seems likely that new technologies will allow markets to evolve that would permit viewers with minority tastes (at least in developed countries) to enjoy material they like, though at some cost.

The international implications of such a development are considerable. In many countries minority ethnic groups, sometimes parts of majority nationalities elsewhere, have previously not been served by broadcast TV. Now they can be served transnationally, as for example in Belgium, where cable systems carry programs from France and the Netherlands to Walloons and Flemish, thus weakening national cohesion. Similarly, cable systems in Northern New England carry Quebec French stations to French-Canadian minorities in the United States, and California cable systems carry Latin American programs.

A second international implication is that it will often be easier to build an adequate market for programs that appeal to unusual tastes by marketing it on a broad international rather than on a narrow national scale. This will lead to a weakening of the territorial and national focus of programs.

Third, national broadcasters will lose audiences. Cable television is often succeeding because of the public's desire to receive foreign broadcasts. The Netherlands, Belgium, Switzerland, and Canada are examples; two out of ten Canadian homes can and do watch U.S. TV over the air, and an additional four out of ten can receive U.S. programs over cable. Canadian broadcasters and Canadian nationalists resent this, but there is no doubt about the public's choice.

Thus the opening of new and enlarged channels for video distribution creates opportunities for an expanded international flow of programming through local redistribution channels.

Person-to-Person Networks

So far I have discussed electronic mass media and how they are becoming capable of delivering material across frontiers to specialized audiences with diverse tastes via current and emerging technology. The media that I turn to now are also likely to become highly multinational, but, because they are new, there is not much history to be reported. I must inevitably treat their consequences somewhat sketchily, because extensive international use of these new technologies lies mostly in the future.

The creation of mass media was one of the great sociotechnical inventions of the past two centuries. By delivering identical messages to thousands and millions of people, it became possible to provide extraordinarily attractive material at pennies per exposure. Slick magazines with colored pictures, thick newspapers with detailed news and features, top musical performances on radio, and exciting drama in movies and TV gave the broad populace a daily exposure to a level of entertainment and information available in the past, if at all, only to a privileged elite.

This is no encomium to the cultural level achieved by the mass media, only to their superb skill in being attractive to their audience. Economy, professionalism, and excitement were obtained at the cost of variety, flexibility, participation, and interaction. Few barbershop quartets, school plays, or conversations have the polish of a grade B movie or soap opera. One recognizes them as awful, from the detached perspective of a media critic, if they are mistreated to a recording. Yet for their participants they have a value that far exceeds even great art. They are theirs; they are self-expression; they are part of the participants' indeterminate growth as people. They are communication rather than reception; they express a democratic activity of the participants rather than a passive acceptance of what is given.

Two-way communication creates less resentment. It is felt less as cultural intrusion and incurs less sense of domination than do the mass media. A phone call is an expression of both sides in a way that a broadcast is not. A search of a library is an expression of the needs and desires of the searcher in a way that a newscast is not. Telecommunications, as contrasted with mass media, can facilitate cultural diversity. In contrast with monolithic mass media which can only be distributed in their fixed forms, interactive two-way media may offer opportunities for cultural tuning, diversity, and multipolarity of origination that do not exist today.

Yet, even if the problem of cultural intrusion—so bothersome to some in relation to the mass media, such as broadcast television—does not arise to the same degree for the new technologies of point-to-point and computer telecommunications, still other concerns remain. There may be some xenophobia. There may be fear by nations that are not on top of the technology that a scientifically sophisticated communications mode will be dominated by those countries where the experts are now located. And there may be a general, often irrational, fear about computers; alarmists often perceive them as magical devices that can somehow control people's lives and subject them to foreign domination.

Telephone

Most familiar in the expansion of international communication is the long-distance telephone call, whose use has grown by leaps and bounds, both for public and private networks.

One dramatic and novel impact of the international telephone system is its use by dissidents in dictatorships. Forbidden to travel, watched when they meet foreign visitors, with no way of knowing for sure what happens to a letter deposited in the mail, they have turned with remarkable effect to the long-distance phone call. That is how the world and distant friends have known each day who has been arrested, what manifestos have been issued, what threats have been received. Some restriction and censorship of telephone calls is possible, of course, but as the volume of international calls and of telephone subscribers increases, an effective monitoring becomes unattainable.

The international telephone network has had a profound but largely undocumented effect on the conduct of international business. In 1963, according to a survey I conducted, international businessmen made surprisingly little use of foreign publications; when they had important decisions to make and needed reliable information, they believed there was no substitute for boarding a plane and going there.[22] In retrospect it seems extraordinary that respondents to my survey did not mention the telephone. But this has changed over time. The fact that international telephone rates have fallen is part of the explanation; for important business calls, cost is not critical, though for routine ones it still is.[23] The most important part of the explanation is the introduction of Direct-Distance Dialing (DDD) and the expansion of the number of circuits so that the chance of completing a call is reasonably good. A jump in usage follows whenever direct dialing is introduced, which is happening around the world as stored program electronic exchanges are being introduced.

The cost of international direct-dialed calls will not necessarily fall at the speed that the technology would justify. Distance, as we have seen, no longer accounts for differences in telecommunication costs. A direct-dialed call in the middle of the night for a 3,000-mile distance in the United States costs only one quarter of a similar call at the same time for about the same distance from the United States to Europe.

The explanation for this differential is partly political. There are some additional costs to international calls because of the need to harmonize systems. But the most important difference is in pricing policy. Most national Post, Telegraph, and Telephone Administrations (PTTs) as a matter of policy try to make as much profit as they can on the international service which is used by businesses and well-off individuals; the extra returns are used to keep costs low for ordinary residential users.

Yet, however great the strategic resistance might be to bringing down international phone rates to true costs, the direction of movement will undoubtedly be toward greater approximation of the two. As costs are further reduced by more advanced transmission, new satellite generations, submarine fiber links, and such techniques as packet transmission, international traffic is bound to grow, even beyond the 1980s rates of 15 or more percent a year.

Data networks

Along with growth in voice communications, there is the equally important growth in international computer data networks. They can be used for data, text, or voice, packet or otherwise, for message exchange or for time-sharing access to information. There are no clear distinctions among the uses since the hardware for all of them is the same. Examples for the new types of networks are Telenet and Tymnet, the first providers of commercial packet service.

For international communication, data and facsimile services that leave a written message have some advantages over voice. Time-zone differences force conversations into inconvenient short windows of a few hours that are working time at both ends; text messages, on the other hand, can sit till people arrive at the office in the morning. Where language difficulties are present, it is often useful to study a text and to puzzle over it with a dictionary, instead of trying to decipher ever louder repetitions of ill-understood speech.

Perhaps the best clue to the impact of low-cost long-distance data communication is what happened to American business and social organization when long-distance telephoning became abundant and reasonable in cost. There was, as we noted above, a redistribution of activities to new centers. Centralization and decentralization coexisted; the size and complexity of organizations grew and so did heterogeneity and dispersion within organizations. Without artificial obstacles it is hard to see why similar impacts would not occur internationally with long-distance data communication.

Computer conferencing

Located somewhere between point-to-point media and mass media are group media. Cable television, for example, can be a group medium if channels are abundant and cheap enough to be used for a mere score of viewers. Telephone, too, can be a group medium, for example, in the "chat-line" services.

Linking several subscribers either through a switchboard or on a party line was a technique available from the earliest days of the telephone. Teleconferences were set up by telephone promoters for demonstration. Where there were party lines in rural areas, sometimes the farmers would meet at a regular time over their telephones. However, the quality of voice on the phone was such that the extensive use of teleconferencing failed to take off for a long time.

Part of the problem with conventional phone equipment for *audio*conferences is the difficulty of identifying who is talking, which limits the number of participants. That is less of a problem with *video*conferences, but these are expensive and still require advance preparations, special rooms, and so forth; before such usage can become widespread, bandwidth must become much cheaper than it is now. What is possible now, and cheaply, is *computer* conferencing, in which written messages are exchanged over a computer network.

Computer conferencing has existed in simple forms only since 1970. It was developed by the U.S. government's Office of Emergency Preparedness as a means of interactive communication between its offices scattered across the nation. The mechanics are fairly simple: The conferees are at dispersed terminals. The computer accepts messages from the terminals; these may be public messages addressed to all participants in the conference or private messages addressed to specific persons. Whenever a participant logs in at his terminal he receives whatever messages have arrived since his last log-in.

There are several advantages of computer conferencing that are not shared by face-to-face meetings: participants can take part from any place where they have access to a terminal. There is no requirement to have all participants logged into the computer simultaneously; they may participate at a time of their own choosing. This temporal freedom is especially useful for international conferences where the participants are in different time zones. The absence of time pressure afforded by computer conferences permits thought-out answers. Conferencing programs exist that can poll participants on their views or votes. The absence of physical interaction also reduces the social pressure toward "groupthink."[24] It also eliminates obstructionism by one member who monopolizes the meeting time by frequent and lengthy contributions. In computer conferences, other members can skip over such material and discourse among themselves. Participants can skip those discussions not of interest to them and read only contributions that are. Also, in a computer conference, everyone can speak at once. When an especially provocative statement engenders response in a live conference, each speaker must wait his turn, and waiting modifies the original spontaneous response. The gist of the discussion may shift to a different topic. In a computer conference, when several participants are on line and a provocative statement appears, each participant can respond immediately at each of the several terminals without being preempted by other responses.

There are of course disadvantages also, most notably the inability to perceive the physical reactions of fellow participants. The tone of voice, grimaces, twitches, arched eyebrows, and other body language of conference participants may convey more meaningful information than their words.

A positive institutional effect of computer conferencing is that it can link invisible colleges—groups of professionals doing similar research. It may permit more participation of academics in smaller universities in such colleges which tend to grow around communication links. But certain aspects of computer conferences can have unsettling effects on existing organizations. Contributions to a conference, appearing only as print on a terminal, are more informal in style and are judged on their content, less influenced by the status of the originators, positions in the hierarchy, sex, age, and so forth. Inputs come in from all levels. The ability to submit anonymous messages allows the raising of an issue that might otherwise remain hidden because no one wants to be the one to bring it up. The use of teleconferencing to coordinate geographically dispersed organizations seems to work to the advantage of outlying regions, which find it gives them more power. Headquarters managers who have hitherto controlled the flow of information are less enthusiastic. Teleconferences subvert hierarchy.[25]

Murray Turoff has studied the cost of computer versus live conferencing and the number of words that may be exchanged among members of a group in a given time.[26] For any group that has a casual typing capability of 30 words per minute, the computer conference becomes more efficient with four or more people in the discussion and halves the time needed to exchange a given number of words when the group has increased in size to ten people.

International Information Retrieval Systems

In 1937 H. G. Wells forecast the encyclopedia of the future; it was not on a computer—he did not anticipate that—but it was to be in microform. "A world Encyclopedia no longer presents itself to a modern imagination as a row of volumes printed and published once and for all, but as a sort of mental clearing house for the mind, a depot where knowledge and ideas are received, sorted, summarized, digested, clarified and compared . . . This Encyclopedia organization need not be concentrated in one place; it might have the form of a network . . . It would constitute the material beginning of a real World Brain."[27] Wells foresaw that the encyclopedia might be a network; in 1936 he could not see how, and could not be sure. Fifty years later we can be.

The first few ganglia of Wells' world brain already exist. Information retrieval has become a big business. The industry is also called electronic publishing, which perhaps describes it better, for it is the dissemination of information in electronic form. The concept electronic publishing also includes publishing done on printed paper but produced by computer, such as, for example, the hard-copy volume of an airline guide.

The large on-line information bases today are in proprietary systems such as airline reservation records, in company inventory systems, or in the financial and personnel records in management information systems. In many companies these are accessible from terminals located all over the world.

The main on-line information bases published for general use are:

> financial information services;
> real estate listings;
> business and consumer credit;
> marketing data;
> economic data;
> legal services;
> bibliographic services;
> scientific and technological information.

As on-line information retrieval grows and takes over many of the present functions of reference publications and libraries, we must think about the changes in relations of work and property that may be created by this new means of information, and also about the institutions and practices likely to emerge from those social relations. And here, most specifically, we are interested in how far that industry is likely to be transnational. Because the industry is, historically speaking, still quite young, present practices provide only limited clues. We must extrapolate from the nature of the technology.

Data creation versus data marketing

The library organizations that disseminate data bases through their computers are not normally the producers of the data bases. That separation is likely to continue. The distinction is the same as that between libraries and print publishers. The economics of the two activities is quite different. Libraries need to be comprehensive. At least within a single field one wants to find in the same location material from any publisher. Publishers, whether for print or computer data, are apt to be large but competitive, since beyond a certain point there are no great economies of scale. Producers of data bases tend to be narrow specialists in the subject matter of the data base. Often they are the authoritative leaders of their chosen field.

The relationship between the data-base compiler and the on-line service operator is still evolving. Some data bases are sold outright; others lease their data for a fixed fee plus a royalty for each citation printed; others ask a percentage of the income; and still others are mounted on line for a fixed fee, with the data base owner doing the marketing and keeping the income.

The borderline between the business of the information compiler and that of the disseminator is a fluid one. There are always territorial conflicts between the communication utility and the businesses that rest on it. Just as Western Union and the press fought over who should run the news agencies, just as AT&T and RCA jockeyed over the relationship of radio broadcasting and programming, so there will always be conflicts and varying solutions to the relationship between the vendor of information services and the compilers of the information that is vended. Yet somewhere the line will be drawn, and those two very different activities will probably remain separate.

This raises the question, Why is there need for a middleman, a vendor at all? Why shouldn't the ultimate customer access the files of each original data compiler via the telecommunications network? There is no technical obstacle. Data compilers are all on the telecommunications network. However, intermediate wholesalers serve at least three functions, and so are likely to continue to exist. First, they provide the user with a guide to what information is available and where it is; this is the classic function of the reference librarian. In the second place, they may interpose a uniform search protocol between the user and the data. No user is going to bother to learn many different search procedures. In the third place, the vendor serves as a unified bill collector, saving the makers of every small data base from having to set up a global credit and collection system for everyone who wants to use their services.

Data bases are a business; they differ from print libraries, virtually all of which are subsidized and most of which are free. For that reason information retrieval has spread much more rapidly among business users than in academia. Universities treat students' and faculty's time as being of no economic value. To spend money on the library to save time for its users does not help the university budget in any way, and therefore not enough attention is paid to making searches efficient. A corporate research unit is in a very different situation. The company pays for the time of its researchers. It is worthwhile to spend money for a retrieval service that will save employee time.

Costs and the structure of the industry

The economics of data-base creation would seem to favor the emergence of plural independent centers, in many instances with a monopoly on their particular product. Putting together a listing of all the physics articles published in the past decade is a massive job, and when someone has done it once, it is not likely that someone else will find it sensible or profitable to do it a second time. Compiling a data base of telephone numbers requires information that often only the phone company has. Putting together an index of the *New York Times* or *Le Monde* or the *Hindustan Times* can be done most efficiently from the electronic record from which each of these papers is composed. A data base on Indian agriculture can be assembled most effectively by the Indian government from its national agricultural survey. Thus, either the first entrant, or producers in particular places and organizations, will have an insuperable advantage in the production of particular data bases and will thereby gain market power. The data-base industry seems likely, therefore, to be a classic case of monopolistic competition.[28] Data base production is by nature a highly dispersed activity, conducted in many countries (and not only developed ones) and language-sensitive once it reaches consumers.

In contrast, the data distribution market is likely to be oligopolistic—unless government regulation turns it into a monopoly. Up to a moderate level there are distinct economies of scale in computer library operation, but not beyond that level.

The costs of furnishing on-line information services may be divided into three categories: (1) creating the information and converting it into machine-readable form; (2) maintaining it on random-access memory with a suitable communications interface to the network; and (3) searching for and transmitting to the customer the information that he requests. Creating and storing the information account for the largest part of the costs by far. Fixed costs are large, while the incremental cost of an additional access is fairly negligible.

High fixed costs and low variable costs, the usual source of economies of scale, characterize most computer operations until a given computer gets to be heavily loaded. The elements of fixed cost in data archive operation are the acquisition of the data archive itself and also the computer hardware through which access is achieved. The data archive is likely to account for the largest proportion of cost. Assembling data is a labor intensive, highly skilled operation whose costs are likely to rise. Furthermore, unless a data library is large it is not likely to be of much use. A large data base also implies a correspondingly large memory in which to store it.

The size of the central processor can vary with the expected volume of user inquiries, but it cannot vary quickly, for it is a major investment. Furthermore, it must be adequate to handle peak loads. As a result, much of the time there will be substantial unused capacity. The variable costs of a search are likely to be quite small; in general the variable costs of computer operations are trivial until the system is being pushed beyond its normal capacity.

Another factor, however, favors a tendency toward global expansion of the geographic base of retrieval services. It is the peak load problem. An optimum solution involves load leveling, which can be facilitated by spreading the service over many time zones via an international network. This is one of the bases for the success of international data retrieval systems, because load leveling more than compensates for the telecommunications costs. Thus there is a strong incentive to spread the market for an on-line computer service both eastward and westward, as far as possible.

The forces for monopoly have limits, however, because high profits attract other entrepreneurs. Given the leader's economies of scale, a challenger will usually offer a slightly different product. Data distributors will be tempted to try to acquire exclusive rights to a variety of data bases, or to create dedicated networks of customers that are not open to other vendors.

From a social point of view there are strong arguments for a universal system not segmented that way. Users do not want to discover that public data is distributed only through some vendor with whom they have no subscriber relationship or who uses a protocol or terminals that they cannot. A universal system is created either by having a single organization or by having compatible standards. But this raises all sorts of policy issues of a sort that are very familiar in the telecommunications area, and to which we shall return in other contexts below. For the moment it is enough to flag the fact that information distribution—unlike its compilation—is that kind of an inherently oligopolistic industry which leads entrepreneurs to play exclusionary games with standards and rates, and which leads regulators to enter the game either to enforce competition by controlling the oligopolists and setting standards, or to protect a monopoly. All these things are likely to happen somewhere.

The world brain that H. G. Wells forecast will include in its computer network both the many small and dispersed creators of data and the smaller number of service vendors. Indeed, it is an operations analysis matter to optimize the frequency and fullness of transmission of data by its compilers to the vendors. Some files, once created, are permanent; for Chemical Abstracts, for example, it would presumably be optimal to store them in the vendor's library files, with the Chemical Society transmitting additions only, and doing so perhaps every day or every week. Other files are constantly changing; current meteorological measurements may not be archived by the vendor at all; it may be optimal simply to put the user on line to the vendor.

As the cost of computer memory falls it becomes economical to store large masses of information at many locations and thus to avoid the communication costs of shipping data from place to place. Yet as communication costs fall it also becomes economical to centralize data and access it from afar. While the cost of computing is clearly falling faster than the cost of communications, the ultimate balance between them is uncertain.

We can predict one thing: the answer to the trade-off issue is that both small computers and centralized computer networks will be with us. To estimate how each may be used at any given time one must consider not merely the costs of storage but to a much greater extent the costs of generating the data, checking it, updating it in the various files in which it may be scattered, and seeing to it that those updated files are consistent in their various parts.[29] Those are expensive activities. One would like to avoid duplicating them in several places, especially for data which undergo frequent changes.

The falling costs of communications over far-flung networks and the falling costs of local storage both contribute to reducing the otherwise growing burden of information management. There is a role, clearly, for both low-cost scattered storage of data and for low-cost transmission of data over networks, with the balance between these options constantly shifting as technology changes their relative costs. They are alternatives with an inverse proportional relationship to each other, and they have important implications for the geographic structure of the industry.

From a technical point of view, a vendor of information services could be headquartered anywhere, or indeed could be dispersed over a network in many places. It is interesting to note that two large firms providing libraries of information on line are both in California. Neither firm thought it important to reduce communications costs by putting their central computer in the middle of the country. The communications costs were not the critical variable. More important was putting their headquarters in a place where skilled computer scientists would like to live.

Whatever the networking arrangements, there are not likely in the end to be a large number of separate physical storage arrangements of mammoth size attempting to be compendia of all the world's knowledge. The situation is not very different from that of book libraries today. With a seven-year doubling time for research publications, only a very few great depository collections like the

Library of Congress, the British Library, or the Moscow State Library can exist. And even those half dozen or so famous libraries are less and less able to cope with the flood of documents, reports, and other fugitive materials that contain much of what is essential knowledge in a period of rapid change. The Harvard University Library and the New York Public Library have made American publishers rather unhappy by taking the lead in forming a cooperative arrangement for division of labor, specialized collecting, and sharing of resources, partly by telecommunications. Those great libraries have recognized that they are reaching the inflection point at which exponential growth of collections of books and documents must slow. Increasingly, they will have to find ways of providing access to knowledge by telecommunication between the users, wherever they are, and the medium storing that knowledge, wherever it happens to be located, rather than by depositing copies of all documents in each library.

All of this implies the end of primary depositories and the retention of much data in its normal place of origination. There is a limit to what can be prepared for library deposit. For those who want the data more quickly, or want detailed data that has not been compiled, the ultimate means in the future for collecting data will be to get on line (with permission) to the operating data (or to an authorized subset thereof) of the source organization. The advance filtering of such data in order to assemble the particular data one thinks will be wanted into static collections in research archives is a massive operation that can only be economically justified for data for which there will be a predictably high volume of use. Wasteful activity in archiving can be reduced if networking permits data to be accessed from its natural habitat, where it accumulates in the process of daily work.

This diseconomy of centralizing all data becomes obvious as soon as one starts designing an actual information system. Among the major kinds of uses of computer communications, many involve distributed origination of data on a continuous basis. In point-to-point message systems, in management of a large organization, and in funds transfer, activity goes on all over, and it would be economically absurd to try to capture all of the data in any one place. Time-sharing systems in the present computer generation do bring all calculations to the host computer, but with the development of microcomputers and intelligent terminals, only the occasional calculations requiring large core or archived data must travel, and most smaller operations can remain localized. The economies of information retrieval from data archives is, as we have noted, sensitive to the outcome of the race between alternative technologies. It is clear, however, that there will be no economic advantage to bringing together all specialized archives under a single roof. The old bugaboo of a national data center into which all data on everyone would be assembled in a single place has vanished, not only because it is a nightmare to libertarians but also because it no longer would have any technical or economic advantage.

Billing

A problem in vending telecommunications services is that the vendor and the customer have no close contact with each other. They need not meet to bargain and "press the flesh." Where the customer is a substantial institution, like a business, that makes little difference. In such cases there are established credit and collection mechanisms. The customer will not disappear. In the case of small consumers operating out of their homes, their offices, or in transit, the problem is quite different. Either the vendor must establish a large sales and collection organization or it must somehow automate collection.

There are various ways to do so. One is to have coin machines, like the copying machines in a library. That, however, requires special equipment to be widely distributed and therefore a large operation. Furthermore, it misses the opportunity for low-cost dissemination created by the presence of TV sets and telephones as terminals. Thus one obvious solution is to turn either the phone company or the cablecaster into the bill collector. The phone system has established, and the cablecasters hope to establish, a widespread system of equipment which reaches every potential customer for information services and which also has a billing and collection organization. The organization that provides the conduit could, for a fee, append to its bill charges on behalf of the service that was delivered over the conduit.

Thus one initially attractive model for the financial transaction system is that of the telephone network where the user deals with only one vendor, the local phone company, although the service received may originate with many other organizations. That is how the customer pays remote and even foreign phone companies for their part of the long-distance calls the user makes. The division of the take is invisible to the customer. The various videotex systems, such as the Teleset system in France or the videotex transmission services of several American Bell companies, uses this gateway billing arrangement.

Another model is that of credit cards. An information vendor, upon receiving a request for service which would include the requestor's credit reference, verifies credit with the referenced institution over the network before providing a service. It is also possible to combine the last two options by having the telephone company forward information about the calling party to the called party for approval.

Another plausible model is based on the assumption that online information services to small consumers will come widely only with the arrival of electronic funds transfer (EFT) systems for the same population. EFT also requires a widespread system with fairly universal acceptance and adequate controls over the behavior of its users. Once that system has reached the individual household, there will be a bill collecting mechanism reaching virtually every customer electronically.

These problems in creating a viable system for billing are even more complex internationally. One can imagine that many vendors will not choose to respond to retrieval requests from countries where they have no established financial arrangements. This provides a role for PTTs acting as middlemen in the process. Just as in the world telephone system, they would do all the collecting within their own countries and then only clear aggregate balances between countries. There are differences, however. A computer bill may be much larger than a phone bill. Also the foreign settlement in the telephone case is with another phone company. In the information services case it may be with many vendors, possibly requiring a U.S. intermediary.

All of these considerations reinforce the role for the middleman data vendor. The international difficulties will perhaps push many vendors to one of two arrangements: either wholesaling through a local intermediary, or subscription arrangements in advance, perhaps for unmetered or liberally metered service. The communications technology does not require local intermediaries. Low-cost direct access across the globe is, as we have seen, increasingly possible. But fiscal considerations may impel vendors to route their information through a locally franchised agent. The alternative of advance subscriptions to relatively unlimited services also corresponds to the economics of the industry, since the variable costs of extra inquiries to the data base is small. Perhaps the door-to-door magazine or encyclopedia sales people of the year 2000 will be selling subscriptions to data-base services. In both cases it makes sense to charge a flat fee in advance. Local agents, however, are the heart of that business.

It is important to keep in mind the global pluralism that seems likely to emerge in data base creation as a corrective to the fantasies that exist about world domination by the big brother computer. That, like all symbolic thinking, is a simplified extrapolation of some facts: the fact of American leadership in this field; the fact of the monopolistic character of telecommunications organizations in almost all countries; and the possible fact of the similarly large character of the prospective vendor organizations with their big computers. But the information base industry itself is another matter entirely.

International implications

How far centralized data bases eventually come to be searched from all over the world, or how far the data base industry becomes decentralized and localized, depends on a number of noneconomic considerations as well. For reasons of pride, national security, or mercantilist policy, nations may establish their own duplicative data bases even when it would be more economical to search bases abroad. Some nations may even prohibit use of foreign data bases to protect the business of data suppliers at home.

Along with crassly political limitations, copyright, too, will cause departures from behavior predictions based upon simple economic calculations. Copyright fosters a temporary monopoly. Its effect, however, is limited in the common case of data that can be had by anyone at the cost of some effort and that can be sold to anyone. Bibliographies are an example; anyone who wishes to do so can create a data base of the authors and titles of published journal articles and sell a search and retrieval service to find them. Copyright may apply to the format of that bibliography, but it cannot prevent another company or another country from doing substantially the same thing using a different format. Archives of unique documents could, however, be made monopolistic internationally by their copyright owner.

This discussion of computer information networks so far has used mostly American examples. They are undoubtedly the most advanced, but there is nothing in what I have so far described that is limited to any one nation. Information systems are budding everywhere. The proprietary ones like that which uses the SITA network for the airlines, or that which uses the SWIFT network for the banks, are undoubtedly the largest, but public information retrieval systems are also being launched. Euronet, supported initially by the European Community, is the largest public system. It is the brainchild of a European Congress on Information Systems and Networks that was held in Luxembourg in May 1975. The congress proposed a data-base cooperative which would share files through a distributed network of computers.

Another issue arises from the European PTT monopoly tradition. It is not inconceivable that the monopoly concept could be extended not only to the carrier but also to the information gateway needed to access data bases, or even to the data library itself. It would be the same as using the fact that the post office is a public monopoly to justify barring access via the mails to foreign libraries, newspapers, and press services. That may seem an inconceivable trend in Western-style democracies, but it is a danger that needs attention.

In the long run, when computer data bases become a main means of dissemination of knowledge, the consequences of such monopolization of the right to supply information would be a tragic departure from Western intellectual traditions. The consequences would be not only ideologically deplorable but also economically damaging. If the existence of networks connecting multiple distributed data sources with multiple distributed data users facilitates the economic allocation of human activities and reduces duplicative activities, then that is just as true across national borders as within them.

While the sheer costs of transmission in the satellite era are likely to be sufficiently distance-insensitive to justify global data networking, there are reasons for regional development of many activities. Political and social affinity between countries makes it easier for them to work together on network development. Language areas are likely to develop their own services. Also, some kinds of data are most interesting to those living in the nearby areas—market and agricultural information, for example. But from a technical point of view, national boundaries are a sheer irrelevancy. If there are good economic arguments for regional division of labor among distributed data sources, and for access to remote sources, then the argument is just as strong internationally as nationally.

One may hope that there will be enough freedom in the communications systems of the satellite and computer age that ordinary people everywhere will be allowed to decide for themselves what retrieval systems they find most convenient and what libraries they wish to use. The technology and the economics will permit it.

Chapter 7

Regulating International Communication

International communication is often considered a mixed blessing by rulers. Usually they want technical progress. They want computers. They want satellites. They want efficiently working telephones. They want television. But at the same time they do not want the ideas that come with them. They lament erosion of cultural integrity. They complain about foreign concepts. They worry about rising expectations. They resent the decline of compliance and of acceptance of the status quo. They object to foreign enterprises cutting in on domestic media and national telephone monopolies. They fear the fiscal costs of competition.

Since rulers are sovereign, they combat these threats by regulations intended to protect their countries' established communications institutions. In authoritarian and totalitarian lands rulers do so unreservedly, for it is policy to control the content and conduct of communication. But they do it also in democracies where befuddlement of policy has accompanied the new communications technologies. The principles of freedom so painfully established for print and oratory have been inconsistently applied to new means of communication.

I have treated these legal changes in *Technologies of Freedom*.[1] Here, I will look first at restrictions that are typically imposed on international communication and at the reasons given for applying them, and then ask what the social sciences can tell us about the consequences of those policies.

Restrictions on Free Flow

Perhaps only Kafka could do justice to the web of regulations that enmesh communication almost anywhere. I shall attempt merely a brief primer here. Virtually everywhere the postal service is a monopoly, and almost everywhere the same is true for telegraph and basic telephone.[2] Whenever the government establishes a monopoly, it is illegal to provide services in competition with it. Leaving a message for a friend is legal, but it is illegal to carry such a message for a third party. This is called the third-party rule, and it falls to lawyers in different countries to worry about the fine lines. The doorman who takes a message and delivers it is probably behaving legally because he does not enter the public streets, but a telephone answering service in many places is breaking the law in passing on the messages that are left with it.

Cable television systems could infringe upon the monopoly, so they cannot be set up without permission. The usual rule allows the wiring of a common antenna within a building, but if the cable crosses the street it is deemed to infringe the telecommunications monopoly.

Walkie-talkies, citizen band radios, and satellite systems may also infringe the monopoly if they permit direct communication between persons and thus bypass the public network. The national post, telegraph, and telephone administrations (PTTs) object, for example, to any satellite system that allows customers to have their own private small earth stations that are linkable via foreign or competing satellites. Only political obligations, not technical requirements, compel the users to be served by a satellite that belongs to their country's telecommunications carrier or affiliated system. Companies and institutions in a country that does not yet have its own satellite might well wish to have direct transmission between small earth stations via a foreign satellite, thus totally bypassing the domestic monopoly. That, however, is usually forbidden. Countries are inclined to require traffic to come into their national ground stations and travel on the domestic system to the customer, even in instances where the country's domestic system is vastly overloaded.

In almost all countries large telecommunications users may lease private lines. The PTTs in some countries hope to abolish them, however, as soon as they can provide an adequate alternative on the switched public service. In most countries a private line cannot be used to switch traffic freely between it and the public network.

In many countries other than the United States it is illegal to connect any unapproved "foreign attachment" to the phone system. Often, devices like telephone answering machines or modems have to be leased from the PTT, though this restriction is being loosened. Even if one can get better or cheaper modems elsewhere, it is not legal to do so. For computer traffic this can be serious, since the quality of the modem may affect the data rate that one can reliably achieve over the line.

In many countries if one wants to attach a switching computer or other computer or terminal or facsimile machine, the rate charged for the same line will be higher because it is transmitting data rather than voice. The rules of the Consultative Committee on International Telegraph and Telephone (CCITT)—the international coordinating body headquartered in Geneva—allow for such a surcharge for data, though there is no technical difference in the transmission stream.[3]

Rates almost everywhere are set politically. In many countries the rates are controlled by a ministry of communications (whatever its name), in others by the parliament, in the United States by state and federal regulatory commissions. Where the rates are set by the ministry, the public has little protection, since the ministry is in effect setting the rates for its own services; under those circumstances there is usually no significant provision for adversary hearings. Where the parliament sets the rates, or the major rates, it is the other way around. The politicians avoid raising rates for their constituents and are likely to force the services into deep deficits before any action is taken. Regulatory commissions, as in the United States, have detailed rate setting as their main task and are therefore likely to get wound up in extremely complex proceedings.

Whatever rate-setting principle is followed, the result is a significant bias in how the system works. In the United States the basic principle followed for a long time is that private but regulated companies should be allowed to earn a fair return on their capital. In most countries the basic principle is so-called value-of-service pricing; a class of customers is charged what the service is worth to them (for example, more for data than for voice) or, in other words, what the (monopolized) market will bear.

Technical standards, as well as methods of pricing, are set by conventions of the CCITT, or in the case of radio by the CCIR (the Consultative Committee on International Radio); these are subordinate bodies of the International Telecommunications Union. Agreement on common standards facilitates interconnection of the various national systems. Outside the United States the CCITT standards are adhered to rather closely. While standards facilitate interconnection, they restrict innovation, for new devices may not fit existing standards. That is an inevitable trade-off. If standards are set too loosely or too late, various devices cannot be used together. (For example, the videotape systems of different manufacturers have been incompatible.) If they are set too soon or too rigidly, they prevent improvements and innovation. Within the broad frame of the CCITT standards, governments sometimes set still narrower, more rigorous and idiosyncratic standards so as to provide protection for their own domestic communications equipment manufacturers. France, for example, chose to use a color TV system (SECAM) different from that of the United States (NTSC) or the rest of Europe (PAL) so as to give their electronics manufacturers a protected market. It worked, but at great expense and trouble in the exchange of television transmission with other European nations. Protecting domestic manufacturers by setting idiosyncratic standards is particularly rampant in the case of telecommunications equipment. Some countries, as a set policy, buy domestic equipment for their communications system; it is hard for governments to resist the political pressure to buy at home. But even when the policy is not explicit, by having their own peculiar standards they make it hard for foreigners to compete. British switching equipment, for example, according to union agreements, had to be raised a certain distance above the floor to reduce the need for bending over. That minor mechanical variation gave British companies an advantage in bidding on exchange equipment. The European Economic Community concluded that there was no common market in telecommunications equipment.

All use of over-the-air spectrum requires government license. Most countries monopolize broadcasting. Those that do not still apply some sort of political or personal criterion when deciding who is to get the scarce licenses. In most countries adversary proceedings on licenses are not common. Countries vary greatly in whether they will allot licenses for person-to-person radio communication. CB systems are rare. Flat prohibitions against listening to foreign broadcasts are less prevalent than social pressure against it. In the Soviet Union listening is not a crime, but repeating foreign propaganda from the broadcasts is.

The United States has a unique set of regulations arising from an intent to restrict monopoly and encourage competition in the telecommunications industry. Since only partial competition exists, companies are understandably tempted to load costs and charges onto the more regulated monopolistic parts of their business and to lower costs and charges where they face competition. To prevent that, government regulatory agencies have set up elaborate rules controlling the methods of accounting and allocating charges.

Canada and Japan illustrate regulations of another type: those intended to protect domestic facilities against foreign competition. For a long time Japan was particularly restrictive on computers. Computer networks were not developed well in Japan, because the third-party rule was applied to prohibit anyone but the phone company from providing a network to a third party's computers. Time-sharing services could lease lines from the phone company so as to link their own computers to customers, but they could not create something like Tymnet or Telenet. But the Japanese policy has changed considerably in the direction of liberalization.

Canada is particularly sensitive to an erosion in control over telecommunication, and understandably so. A significant proportion of its population lives within one hundred miles of the long border with a country which leads world development in both mass media and telecommunications and which has a population ten times as large as its own. It is not surprising that the bulk of Canadian communications are at least partly of U.S. origin. U.S. radio and television can be heard from across the border, and cablecasters carry it to where it would not reach otherwise. Movies, TV programs, records, and magazines are imported in large numbers and have led to restrictive regulations.

If the regulations governing international communication were not Kafkaesque enough, there are, of course, many other regulations relating to interference, safety, taxes, financial responsibility, and so forth. Beyond all of these, there is censorship in many countries and prohibitions against foreign propaganda or against disclosure of state secrets, which can be defined quite widely. I will skip over these rules, important as they are, because our

subject is the inadvertent effects of the normal policies of democracies. In all democratic states, even those that sedulously avoid
any regulation of print media, the regulations on electronic communication are profuse; for the international communicator what
is particularly bothersome is that they are different and constantly
changing from country to country.

International restrictions

Along with the various restrictions on the free flow of communication between nations that are imposed by one country or another,
certain international agreements and practices of a restrictive character are evolving, too. While they are not very binding, the trend
is significant, for over time it has its influence on domestic law.

International law and custom in every era reflects the ways that
nations relate to each other. Strictly speaking, international law,
which evolved along with great nation states, has no substantive
content but deals only with the ways in which sovereign states
interact and reach agreements with one another; what they reach
agreements *about* is not the concern of international law. But one
form of interaction of states goes to the heart of their very existence, and so international law has always been the "law of war
and peace," to use the title that Grotius gave his classic formula-
tion of it in 1625.

As commerce rose in Europe and began to expand beyond, the
law of sovereign states (public international law) began to be supplemented by a whole new branch, private international law, to
cope with the conflicts that individuals got into when trading
across boundaries. Along with principles of law there evolved a
wide range of customary practices that merchants and seamen
could count on as they wandered around the so-called civilized
world, and which to some extent they imposed on ports in the
remote corners of the globe. "Treaty ports" were established as
enclaves of European practice.

Law and custom regarding international communication appear
only infrequently in the classic treatments of either public or private international law or practice. What does exist is mainly concerned with the rights of ambassadors or travelers to carry on their
business. There is an age-old recognition of the right of ambassadors to report back home, unmolested. Beyond that, writers on

the practice of international relations became conscious of problems of international communication only subsequent to the widespread establishment of the post and telegraph.

The International Telegraph Union (ITU) was formed in 1863 when use of telegraphy was but two decades old. Even at that early date the problems of interconnecting the wires at frontiers, of cooperation in laying cables, and of arranging payment for international services required some forms of cooperation. With the coming of radio and radio interference, the importance of the ITU grew, and its name ultimately changed to International Telecommunication Union.

Ironically (but also characteristically), the new medium, telegraphy, set a model for the older medium of the past. A Universal Postal Union (UPU) was formed in 1875 by the Berne Treaty. As has been happening often since, a need for a high degree of organization and regulation was first noted in regard to the new, more rapid electronic media; the slower print media had not required it, and indeed had been protected from it in free political systems. But then eventually the electronic model of organization was reflected back onto the slower print media.

The UPU's primary function was to create a clearing-house system for international mails. Instead of trying to account for and collect individual charges for each letter that crossed a frontier, each country's post office agreed to receive and deliver the stamped mail for which another country had been paid. Periodically, total balances would be struck and payments made through the UPU.

The more difficult problems were those of the ITU. Gradually a body of agreements and practices was formed for the conduct of international radio and for transmission to ships at sea and planes in the air. The allocation of radio spectrum to different services in different zones is done at the World Administrative Radio Conferences (WARCs). The assignment of the allocated frequencies is established by registration with the International Frequency Registration Board. Within any given allocation, the system is first come, first served; whoever wants to register the use of a previously unregistered frequency may do so. This results in prompt use of desirable frequencies and also in the gathering up of them by the advanced countries.

Nature's disdain for national boundaries has led to many of the same sorts of problems for satellite location that exist for spectrum. The geosynchronous orbit, as we noted in Chapter 3, is a finite resource of interest to many nations. Desirable locations on it are the object of conflicting claims. The WARC of 1977 created a regime for orbital allocations that partly resembled the regime for spectrum and partly not. The underdeveloped countries wanted a scheme of national assignment of orbital positions, whether the country uses them or not, so as to assure themselves of the availability of those positions at a time in the future when they may have the resources to use them. The United States objected to the inefficiency of a scheme that leaves many desirable locations idle for an undetermined time. A compromise that we shall describe below was finally struck: different arrangements for the Eastern and Western hemispheres.

Satellites have become a significant focus of international agreements ever since the Treaty on the Peaceful Use of Outer Space in 1967.[4] The details are new, but the basic issues are the classic ones of the limits of sovereignty and the rights of nations against injury by others.

Accepted customs have also emerged about several other aspects of communications, including the treatment of foreign correspondents during both war and peace, practices of tourism, and international scientific cooperation. A nation will be criticized by parties who consider themselves aggrieved if it does not make appropriate arrangements for foreign tourists to travel, drive cars, or buy newspapers from their homeland, or if it prevents its scientists from collaborating with foreign colleagues. Such practices are gradually entering the realm of what is expected of all civilized nations.

A great leap forward in the international agreement on communications occurred at the end of World War II. In that heyday of democratic idealism, after the defeat of fascism, nations widely recognized that it was a human right to communicate freely across national frontiers. A United Nations Educational, Scientific and Cultural Organization was formed as part of the United Nations family of organizations. In 1948 the United Nations General Assembly adopted a Universal Declaration of Human Rights, of which Article 19 said: "Everyone has the right to freedom of . . . expression; this right includes freedom . . . to seek, receive, and impart information and ideas through any media and regardless of frontiers." The declaration, though passed unanimously, only came to be ratified by the necessary states in 1977; up to that time it was a testimony to a general sentiment of the international community, but it was not an enforceable law.[5]

While the positions taken by countries at the United Nations have changed, communication remains a subject of attention. There is a growing attempt to codify international law about it, though instead of codifying the human right to communicate, the majority have attempted to codify the sovereign right of nations to control communications of foreign origin. In UNESCO and the United Nations there have been proposals to formally declare "states responsible for the activities in the international sphere of all mass media under their jurisdiction";[6] to ban militarist, racist, and commercial propaganda across frontiers;[7] to encourage monopoly state-owned news agencies in each country;[8] and to bar direct broadcasts from satellites to countries that do not consent to them. In Chapter 8 we shall examine the last of those proposals as a case study of the way these various proposals for expanded international communications law have been discussed.

So far no repressive proposal has been adopted, despite the fact that some have had majority support. U.S. resistance has blocked action. The United States had to go so far as to leave UNESCO and stop paying its dues for several years. Nonetheless, worldwide concern about issues of communications policy continues to grow, and the pressure for formulating international law on the subject grows, too. In the end, whether liberal or not, communications law is certainly becoming an important new chapter in the body of international law.

The Protectionist Case

SWIFT would lose them 20 million dollars a year in revenue, claimed the American International Record Carriers (IRCs) in opposing that interbank clearing network. That amount—if the figure is to be believed—was what the banks would have saved by exchanging their transatlantic clearings (or the U.S. half of them) on a computer network instead of the way they used to do it. From a social perspective, that is progress; from the IRCs point of view it is a threat. A candid and most common argument of the carriers for restrictions on such efficient new technologies is their effects on present jobs and investments. When organizations such as the post office run in the red and yet employ hundreds of thousands of employees, executives cannot be indifferent to those who lose from electronic competition.

Large institutions with enormous commitment seek always to preserve the right to change slowly. The case they plead is for measured change so that jobs, plant, and service to the public will not suffer. The most common and honest arguments for resisting electronic innovation is one for suboptimization—the protection of a particular set of institutions. But red ink is, after all, society's mechanism for moving resources to where they are needed and confining the Parkinsonism of institutions. To be persuasive, arguments for restriction have to be couched in broader terms, too. Often the argument for protection of the existing carriers gets raised from a pragmatic one about the costs and benefits of change into an ideological one.

Since the nineteenth century the post office has been at the heart of the debate between believers in private enterprise and believers in nationalization. A bibliography of the nontechnical literature on telecommunications policy prior to World War I should show that the references on nationalization versus private enterprise exceed all other topics combined, by two to four times. The argument against nationalization was usually that of efficiency, although socialists tried to use the example of the post office as evidence that public enterprises could be efficient, too. The argument for nationalization was generally one of equity—a case for getting an essential service for the people out of the hands of self-serving entrepreneurs.

The issue started with the mails. It soon extended to telegraphy, which the post office took over in most countries. On more than one occasion postmasters general in the United States sought to get into telegraphy, but it was left to private enterprise. In the years before World War I, the argument spread to the telephone, which was nationalized in most countries but remained private (except briefly in 1918) in the United States.

By now the original socialist arguments have been forgotten by the conservative civil service bureaucracies that run the PTTs, but the sacred character of the national monopoly has remained an article of faith. The case now is apt to be phrased that communication is such a vital element of the life of the country that it, like defense and education, must remain under government control. It is argued further that as a "natural" monopoly it should not be allowed in private hands, and certainly not in foreign hands. It is claimed also that competition would result in less universal service and would favor business and special interests rather than serving the ordinary citizen. Whatever the arguments, the fact is that many PTTs are opposing any system of private message delivery that competes with their services.

If the case for restricting international communication were simply couched in terms of the interest of the carriers, it would not hold very long against the important and powerful interests of people in communication. What tips the balance against freedom of communication is the appeal of nationalism. Three common arguments for regulating international communication claim to be in the interest of the whole nation, not just the communications industry. These are: (1) the classic mercantilist case that exports are good for the country while imports are bad, (2) concerns about sovereignty, and (3) concern for the integrity of the national culture.

Mercantilism

The flow of information is an item of growing importance in world commerce; information is a commodity that is exported and imported and paid for. Information also affects the balance of payments by being the vehicle for technology transfer. Third, the hardware of the new information technologies is among the most dynamic commodities in world trade—color television sets, radios, earth stations, computers, and transistors.

Authorities in many countries worry that with international computer networks and distance-insensitive costs, advanced computing centers (often in the United States) will siphon off service-bureau-type work from their domestic computers. If it makes no difference whether the computer that processes one's work is around the corner or around the world, there will be foreign exchange payments for those who do their computing elsewhere and perhaps less opportunity for them to build up their computing expertise domestically.

The protectionist response is in the form of either flatly prohibiting the buying of computer services abroad or imposing a tariff—

in the case of communications, by charging high rates on the international lines. There are, of course, a variety of motives for these high rates, including the philosophy of charging what the market will bear. But one of the motives is to prevent the growth of international value-added services that compete for business with local ones.

There is reason to believe that such protectionist tariffs hurt those who impose them. The basic argument is simple and is the same as that against protectionism in general: The extra charges that are imposed are a cost of doing business in that country. Expensive communication makes a country less competitive than it would be if it used the best services, wherever they are to be found, to carry on its productive activity. The issue is complex, however, and I will devote a whole section to it below.

Even if it is admitted that not much is gained by a country in using backward information facilities at home rather than sophisticated ones abroad, a complex line of argument nevertheless attempts to justify doing that as a way to sustain hardware development at home. The growth of an electronics manufacturing industry, it is argued, will be favored by using domestic service organizations that in turn are obliged to use home-manufactured equipment.

The electronics industry consists of many sectors, each of which is quite different in the character of its organization and market. There are telecommunications services, computing, and radio and television broadcasting. Associated with each of those services is manufacturing: telecommunications terminals, switches, computers, broadcast equipment, and so forth. There is also manufacturing of components and parts, such as microchips, that are used in several other sectors.

A striking contrast between telecommunications equipment and other electronics can be seen in trading practices. In the former, most purchases are made by the national monopoly carriers, and there is much less foreign trade in those countries that have the industrial capability to make their own hardware.[9] In contrast, for other electronic equipment (which is bought by price-conscious business firms or ordinary consumers), the amount of foreign trade is far larger.

The basic pattern has been described by Raymond Vernon and Robert Stobaugh as a product cycle.[10] At the beginning of the cycle, new sophisticated products are manufactured in the innovating country (usually a developed country). These products yield high prices and profits to their manufacturers, until imitators subject them to competition. With time, as the production process for the new product comes to be well understood and standardized, it is undertaken in countries with low labor costs or more efficient production markets.[11]

Clearly, for areas with high living standards and labor costs, such as Europe and North America, the maintenance of a position in electronics manufacturing depends above all on continued technological innovation. Restricting the use of sophisticated electronic equipment through protectionist regulations assures only that the products in demand will be the standardized ones at the tail of the product cycle, and those products will be made in countries where labor costs are low. If developed countries are to maintain their own manufacturing capability, they must encourage rather than discourage the use of the highest technologies.

Sovereignty

Some nations fear that international computer communication networks can defeat their sovereign policies by providing data sanctuaries abroad. In many fields of international activity there are such sanctuaries; people move activities of a controversial nature to places where they can be carried out without interference. People go from countries where they fear persecution to countries where they are safe. Political exiles from dictatorships publish their journals in democracies, where they are tolerated. Women go to have abortions where it is legal. Rich people put their money in countries where it is not taxed or in Swiss bank accounts where it is not reported. So too with data, one can conceive of motives for keeping it in one country rather than another.

Few people would want a world where dissidents could not escape the more erratic whims and impositions of governments by moving an activity to another jurisdiction where it is legal. This becomes a problem to authorities only when the mobility is on a massive scale or of a very threatening kind. There seems to be no such threat from the flow of data, though two are occasionally cited. One of these is exploitation of earth resource data. In the United Nations, the specter has been raised that a few countries have the capability of observing by satellite the resources of other lands and then using that information to bid for development concessions or to bargain on the world food market. While the gains would usually fall far short of threatening sovereignty, there are clear possibilities of taking advantage of such knowledge. That is why the United States adopted a policy of total public access to all earth satellite data. The problem, therefore, is not the right of access to the information but the fact that many of the countries scanned are not prepared technologically to receive and analyze all the data they want rapidly or completely enough to use it well. A high-speed international data network would help them, not hurt them.

The other bugaboo often raised is that large multinational corporations might keep data on some national market or on operations in that market in a different country to prevent the government of the first country from regulating them effectively. Occasionally a company will undoubtedly spirit out an embarrassing piece of paper to another jurisdiction, just as it may sometimes shred one. But the notion that data can systematically be denied to the host country by a multinational company to the point where the sovereignty of the host is threatened would seem quite topsy turvy. A multinational does business in a host country by the permission and at the sufferance of the host. Nothing stops any government from laying down any levy it wants on data that must be reported, nor is it stopped from requiring documentation of the sources of the data if it so wishes. The multinational has little choice but to comply if it is to stay and do business. It makes little difference in which jurisdiction the data are located. They must be produced. The presence of international computer communication would help rather than hinder the government in compelling the production of the desired data.

On the other hand a data sanctuary can present problems to even firmly resolved governments if, for example, it is a haven against the type of privacy laws that many governments have adopted. Restrictions on the kind of information a credit bureau may keep on an individual could result in the escape of the bureau to a more permissive jurisdiction.

Cultural integrity

The attraction of soap operas, sit-coms, rock groups, and other diversions that the mass media have to offer is not limited to any one culture. These products have a universal human appeal, and that is why a world trade in them has developed. The countries that took the lead in producing pop culture, most notably the United States, during the fifties and sixties became exporters of that culture to the world. While the lead is temporary and fragile, still it has lasted long enough to cause intense distress to those who see in this foreign material a threat to indigenous cultures that they cherish. The issue of what to do about cultural intrusion by foreign material is a thorny one. It pits traditional cultural values squarely against the liberal principle of freedom of information.

Of the more than 150 member states of the United Nations, only a small number may be readily defined as democracies. The majority are dictatorships that restrict their citizens' access to foreign ideas; they limit the books, plays, programs, or information bases that circulate within their countries. It comes as no surprise that in the United Nations they would consistently vote for resolutions to control the international flow of information.

More of a puzzle, and therefore more interesting, is the support these conventional dictatorships have found among intellectuals. Using a theory of cultural imperialism, even some democratic intellectuals have denounced the free flow of information. By that flow, they argue, the dominant powers, most notably the United States, impose a culture of commercialism, pornography, and violence on the rest of the world. It thereby destroys "authentic" indigenous cultures in weak nations and makes them dependent. Moreover, it has been charged that the so-called free flow of information doctrine "is largely a pernicious doctrine, for it masks ideological persuasion and monopoly in an argument of free speech and artistic creation."[12]

Any one such quotation inevitably distorts the views of other individuals in that panoply of overlapping viewpoints which share only their condemnation of the free flow of culture and information. Among them are traditionalists who seek to protect ancient folkways; nationalist modernizers who want to break the hold of a tribal or feudal past; and conformists opposing the introduction of dissident ideas. There are, furthermore, opponents of freedom who see the goals of industrialization and modernization as outside impositions detrimental to the receivers. Others argue for restrictions, but for the opposite reason: that the kitsch which dominates international communications displaces the all-important information regarding development to which poor countries should devote all their resources in order to grow and modernize rapidly. There are those who want to protect the distinctive communications of tribal, ethnic, and religious subgroups, and those who wish to restrict the flow of information so as to suppress factions and to create a single national culture. As stated by Elihu Katz:

> The call for more indigenous creativity on the part of those committed to modernization poses . . . serious questions . . . which tradition should be chosen? There are Zoroastrian and Islamic elements in Iranian culture, Indian and Spanish elements in Peruvian culture, French and Arab elements in Algerian culture—which should be emphasized? The choice, of course, is related to the ideological and developmental goals of the choosers. But even within a tradition, there are problems of selectivity. Certain elements are more compatible with modernizing values than others; indeed, certain elements of the religious tradition are obviously antithetical to modernization. Champions of pluralism and freedom of expression in their "epochalism" may be no less committed to the search for a national identity than the "essentialists." But each group is fearful, justifiably, of the other.[13]

There are also those, and they are the majority, who mix together bits and pieces of these various views.

What those who wish to restrict the flow of foreign ideas have in common is the conviction that the free flow of international communication will somehow obstruct the realization of their own image of ideal national development. And we must recognize that international communication does tend to erode the integrity of established national cultures. Some will see it as being for better and some for worse.

Should South Africa be protected from Tanzanian broadcasts against apartheid? Should Soviet Jews be able to listen to Kol Israel? Should scientific knowledge be made available to schools and universities in developing countries? Should Africans be subjected to Christian or Moslem missionary efforts? Should UNESCO promote cultural exchanges and contact among intellectuals? Should advertisers be able to promote their products in foreign countries? Should X-rated movies be exported to countries holding to traditional sexual mores? Should modern art be available in galleries in all countries? Few people would answer all those value questions on the same side. But whatever one's norms, the fact in the modern world is that there is massive cultural intrusion.

The influence of foreign ideas is but a special case of the disruptive impact of intellect and culture in general. All through history intellectuals have been subversive and have had their products attacked as assaults upon the established culture. Socrates was called "an evil-doer, and a curious person, who searches into things under earth and in heaven, and he makes the worse appear the better cause, and he teaches the aforesaid doctrines to others" (*Apology* 19). And in answer he described himself as "a sort of gadfly, given to the state by God" (*Apology* 30). For this he had to drink the hemlock.

Poetry, philosophy, literature, and the media are perennially tagged as being immoral corrupters of youth, disrespectful of tradition. Those charges have an element of truth. Intellectuals are gadflys. They claim a right to seek the truth and evaluate the good by some light of their own other than the writ of established authorities. They are also, and always have been, conduits for foreign ideas. From the days of Greek slave tutors in Rome, through the wandering bards and scholars of the Middle Ages, to the refugee or "brain drain" scholars of today, communicators and men of intellect have been cosmopolitans and introducers of alien notions.

But something has changed. Phenomena that have been with us from time immemorial are magnified by the emergence of telecommunications and the mass media. That is a quantitative change so great as to become a qualitative change, too. The mass media and electronic communications obliterate many of the impediments that served in the past to slow down changes in ideas and mores. Barriers of time and space that once protected the status quo are easily penetrated or jumped over by modern media. These eliminate the cushion of time between when an event happens and when it is known worldwide. The people have become ringside observers to dramatic news events, be they wars, moonlandings, hijackings, or riots; the public follows them hour by hour while they are going on and while the outcome is still unknown. So too the barrier of distance is gone. We watch an event unfold on a worldwide stage, on which what happens in Singapore is as visible as what happens in Rotterdam or Caracas. Distance is not per se a significant factor, though there still is a vast difference in how we view something foreign or domestic.

Electronic communication and mass media also serve to widen the units in which political, economic, or social action takes place. Peasants who sell their crops in the local marketplace may ultimately be linked to the world market by a series of trading intermediaries, but that is quite different from the situation of the farmer who receives world price quotations on the morning newscast. The latter operates directly in a regional or national if not world market. The village headman whose legitimacy was established by virtue of ties to local lineages and shrines operates in a very different way from the national politician who has to formulate issues as abstract platform planks that will be meaningful in national journals or national broadcasts.

Also, the new mass media operate as alternative sources of information or belief; they create counterweights to the established authorities. Simultaneous radio coverage of a war or moonwalk absorbs and fascinates the mass audience directly, cutting out the traditional local purveyors of information and interpretation. It is not the imam or the chief of state who tells the people what happened and what it means. The people were there, along with the television news camera crew. So, too, the broadening of the arena of action transfers authority from the village bigwig to national leaders, and eventually beyond them to world figures.

In the 1950s Daniel Lerner in a classic study of communication and development, described the difference between three types of persons and three orientations found in the Middle East.[14] Illiterate villagers he described as *traditional.* The world they could understand stopped at the limits of their firsthand experience. They could not conceive of what national politics was all about. Recent migrants from country to city he described as *transitional.* They relied heavily on radio even in that pretransistor era, but they selected domestic broadcasts with familiar themes. They knew about national politics but only as it affected them. A good ruler was one who got them jobs or repaired their town. Educated urbanites Lerner described as *modern.* For them foreign radio was a major medium. They understood the issues of the Cold War and of ideological movements and they followed world affairs with interest. World media enabled them to raise the scope of their empathy well beyond their personal experiences.

A common worldwide focus on news is, however, only one kind of homogenization which exercises those who are alarmed about the violation of national cultures. The examples they cite even more are movies, television, literature, songs, hairstyles, clothing styles, patterns of respect and etiquette, and religious observances.

Popular songs are a common issue between the upholders of tradition and its violators. Stylized traditional music tends to give way to more popular modes such as jazz, rock, or filmsongs conveyed on radio, records, or movies. In India, to preserve traditional musical forms the All India Radio put quotas on filmsongs, the most popular program item. Dance manifests similar tendencies toward cultural assimilation. It is generally a stylized, controlled indulgence in tabooed behavior, whether in boy–girl relations or in expression of other passions. So new-style dances, imported from or influenced by the West, express feelings and break taboos in ways that are bound to shock many who see them.

The same can be said about dress; it symbolizes one's identity. To refuse to wear one's proper tribal or caste dress, or to assume the dress of a foreign elite, or to bare one's legs by wearing a miniskirt is to express a challenging assertion of changing identity.

Eating habits, manner of address, and place of residence all often become highly charged items in the conflict of generations. In most countries the principal battle between traditional culture and modernity is fought over the independence of the young from parental authority. The battle most often reaches its zenith on the issue of arranged marriages and the locus of habitat of the new household, but skirmishes are fought over every sign of self-assertion by the young: impoliteness, staying out late, wearing their hair oddly, drinking Coca-Cola, reading "bad" books, or what not. The central issue is the power of the old versus the autonomy of the young. Nor is the fight just an ego trip by either. In a traditional economy the social security of the whole family and the maintenance of its elder members may depend upon the carrying out of family obligations by the next generation. At the same time, the chance for the young to get ahead depends upon their being able to free themselves from such family obligations. These issues cut deeply and are keenly felt. In Japan, under the Occupation, virtually every novel, soap opera, or movie dealt with the clash between filial piety and the liberation of youth. That, not defeat in a war, was people's central concern. So too, in many countries the clash of generations is the most affect-laden of conflicts.

Religion is another area in which the conflict between tradition and innovation gets expressed. Piety is the most common rationale offered in defending established relations of respect. On the other side, reformations, skepticism, or militant atheism are common attacks upon the traditional culture.

New tastes created in the process of social change provide an economic opportunity to merchants. Commerce seizes upon the chance to stimulate tastes toward what it can sell and to produce for sale what people will buy. Thus, advertising and the monetized economy become major agents of the process of change and serve to undermine family-centered traditional culture. The marketer is a willing panderer to the tastes of transitional men moving naively into a cash economy. Merchants who sell what is popular among those new strata are in turn attacked by traditional authorities as unscrupulous corrupters who will do anything for a price.

These are examples of the areas in which the fight over cultural integrity most often takes place. The media in which novel, expressive, emotionally laden matter that undermines the traditional culture most often appears are art, drama, literature, reform religions, and radical ideologies. All of these appear in movies, soap operas, songs, television, advertisements, pamphlets, and evangelical tracts. It is such media when they come from abroad, containing controversial messages, that most upset the defenders of the integrity of national culture.

The problem of cultural intrusion also arises for the new technologies of point-to-point and computer telecommunications, though in a different way. Where communication is one-on-one it has much more the character of a dialogue, with both partners choosing to engage and acting as initiators as well as receivers. Critics seem less upset about international telephone calls than about the international flow of news, or the international flow of movies, or direct satellite broadcasting, for a very clear reason. The messages of the mass media come in canned form; they are produced in one country and then distributed elsewhere with minimal or no adaptation to the cultural values or local needs of the receiver. Retrieval of data or a telephone conversation, on the other hand, is an interactive, two-way process. Both parties want it; both provide input; each adapts to the other. It is cultural interaction, not cultural imperialism.

Interactive computer inquiries are somewhat like phone calls in that respect. An information retrieval request is initiated by the receiver. Input, too, is more nearly symmetrical than in the case of the mass media. Any research unit anywhere makes its contribution to the world's knowledge. Nonetheless, there are imbalances that cause discontent and lead to attacks on computer networks, telephone systems, and other person-to-person technologies.

A number of fears have been expressed. There is a fear that the world faces a growing gap between the rich nations and poor ones, and between the data-rich and the data-poor. The balance of trade, it is said, will flow increasingly in favor of the data-rich. The sovereign control of their own destiny, it is feared, will be impinged if data operators, particularly multinational corporations, can create data sanctuaries abroad. So despite differences of degree, the issues of cultural imperialism have been raised for all the new modes of communication and need to be examined.

Charges of Cultural Imperialism from the Left and Right

National cultures, when they are lauded by their eulogists, are generally described as age-old traditions. To some degree that is true; every culture has a long history. But to a very large degree, the claims to a hoary past are mythology. Many elements that are valued as indigenous culture were controversial foreign imports a generation or two before. Each generation sees as its culture those values and practices with which it grew up. Its hallowed traditions are those it learned in childhood.

The wish to freeze change at one's youth is a normal, natural human impulse. Few of us learn very much once we are adults. There are few things more difficult than having to alter one's values, habits, and style of life once one has reached the years of maturity. Change, of course, happens to millions of adults, but we must respect the pain it puts each individual through. Old residents of a neighborhood have difficulty accepting an influx of migrants. Members of an aristocracy or plutocracy or ethnic ruling group do not happily accept the rise of equality of those who used to be their menials. Refugees and immigrants, peasants who have lost their land, villagers who have moved to the city, workers in trades that have become obsolete—their unhappiness is one of the commonest and yet most poignant kinds of human tragedy.

It is out of resistance to this kind of tragic change that people become so emotionally involved in preserving their culture as they came to know it early in life. People normally form a sense of identity only once in a lifetime, and they quite understandably become ardently attached to the symbols of that identity.

Media messages march to a different drummer, however. No biological fact slows the processes of media change so as to conform to the human life cycle. The tempo of technological change increases exponentially. The time span of fads grows shorter; the life of popular songs grows shorter; the time on top for a movie star grows shorter; the duration of newsworthy events as measured by the daily newspaper headlines grows shorter. The price of such rapid change is discomfort for individuals whose life cycles move along at a pace no different from those of people who lived and died millennia ago. The drumbeat of new fads, new ideas, new styles every year brought forward by the mass media is in inevitable disharmony with the concept of a national culture and its slow rate of change. A natural response is to attempt to restrict the inflow of new ideas and to attack those powerful institutions which introduce them and prove hard to control.

The literature on cultural imperialism is enormous. Ironically, but not surprisingly, a large part of it is American. That literature itself is perhaps an example of how American writings, representing the biases of particular partisan tendencies in American politics, come, in this era of international communication, to be assimilated abroad in places where their reference is partly irrelevant and where few people have the appropriate experiences to evaluate them. By their very success in disseminating half-truths these authors demonstrate that there is at least a half-truth to what they are saying. American populism with its hostility to big business, and also the condescension of American intellectuals to the commercial popular culture of their country, both get disseminated to the world by the very commercial system of media distribution that they criticize.

Perhaps the most influential book decrying cultural imperialism is Herbert I. Schiller's *Mass Communications and American Empire*.[15] The essential critical ideas are, of course, not new. They are the standard intellectual critique of commercial mass culture, wedded to the standard critique of alien influences, embellished by the Leninist theory of imperialism. That mixture can be found in hundreds of tracts, essays, and speeches.[16]

A thoughtful statement of the concern is that of Elihu Katz:

> Modernization brings in its wake a standardization and secularization of culture, such that the traditional values and arts—those that give a culture its character—are overwhelmed by the influx of Western popular culture. Rock music and comic books and Kojak threaten not only local tribal cultures but the great traditions of societies such as Thailand, Israel and Iran.[17]

More typical is the comment of Guyana's Minister of State Christopher Nascimento:

> How many times greater must be the responsibilities of the media in developing countries than in affluent established societies. How much greater the danger and the threat to development and independence when huge mass media multinational empires own and control the media in, and strive through the use of their vast financial resources and immense international influence, to exercise and maintain monopolies, in developing nations.
>
> It is the presence and the power of foreign owned multi-national mass media empires in developing nations that poses one of the greatest threats to independent development.[18]

Finland's long-time President Urho Kekkonen in a major speech said: "I have read a calculation that two-thirds of the communication disseminated throughout the World originate in one way or another in the United States." A very dubious statistic! He continued:

> These facts make one question the principles of freedom of communication in just the same way as one has cause to re-evaluate the concept of freedom of speech. Could it be that the prophets who preach unhindered communication are not concerned with equality between nations, but are on the side of the stronger and wealthier? My observations would indicate that the United Nations, and its educational, scientific and cultural organisation UNESCO have in the last few years reduced

their declarations on behalf of an abstract freedom of speech. Instead, they have moved in the direction of playing down the lack of balance in international communications.[19]

The international symposium of scholars at which President Kekkonen talked adopted a resolution asserting:

> Each nation has the right and duty to determine its own cultural destiny within this more balanced flow of information within and among nations. It is the responsibility of the world community and the obligation of media institutions to ensure that this right is respected.[20]

Nationalism is not a monopoly of either the right or the left. Rather, nationalism is the doctrine of the right-wing that most easily co-opts the left. Historically, liberals and radicals have been internationalists. Marx's statement, "The workers have no country," epitomizes that view.[21] Liberal intellectuals have fought for freedom of movement, freedom from censorship, and world cultural exchange, and have condemned ethnocentrism and prejudice. Right-wing nationalists, on the other hand, have glorified the unique heritage of their own ethnic group. The right has fought foreign influences that would undermine their historical religion, language, customs, or politics.

But the description of the left as open and internationalist and the right as closed and nationalist is misleadingly simple. Nationalism has always been the most popularly appealing element in right-wing doctrine. As such it has seduced and been adopted by the left. Some leftists got caught up in it; others tried to capture the political appeal of ethnicity by combining nationalist doctrines with their own populist or socialist ones. Before 1917, for example, the Russian and East European social-democratic movement was more driven by the nationalities question than by any other issue. Nationalist-socialist movements of assorted varieties, such as the Bundists, presented a challenge to antinationalists such as the Bolsheviks. Out of the conflict emerged various views on the national question, of which Lenin's are among the most interesting. Lenin berated the "reactionary" character of ethnic identity and at the same time recognized the intensity of its political grip. He understood that the Bolsheviks could not confront nationalism but had to use it. While privately contemptuous of ethnic appeals, Bolsheviks saw them as expedient to stimulate the struggle of subordinated nationalities, in the expectation that the national sentiments of the latter would pit them against the dominant imperial powers.

If the basically right-wing sentiment of protection of ethnic identity has become so wound up in left-wing politics in the past, it is not surprising that the same thing is happening again today. It would be wrong to assume that the people who write about cultural imperialism and warn against violation of the cultural identity of peoples are engaged in a cynical manipulation of a popular slogan to actually implement its opposite. The people who are raising this issue today are generally sincere. It would be more appropriate to remark with astonishment how widely people of the sincere left have been absorbed in classical doctrines of conservatism as a consequence of nationalism. Increasingly they find themselves pushed by the logic of their position to see themselves as more opposed to liberalism than to traditional culture, to assert that the free flow of information is a goal that must be reconsidered, to conclude that preservation of traditional ways is more important than a rapid rise in the GNP, and to justify the use of state power to control the communications that reach the people.[22] Thus a strange alliance puts conservative military regimes or theocratic oligarchies at one with nominal progressives in defense of censorship. Needless to say, what they want censored, what they want controlled, and who they want to exercise that control are quite different. But they unite in advocating restrictions on broadcasting and other instrumentalities of the free flow of information.

Attacks on cultural intrusion may come from any political direction and may serve as a means to any one of a large number of inconsistent political aims. The slogan of protection of national culture most often really means the protection of an existing government, or of some special interest within it. It may be used to protect a local communications industry from stronger foreign competitors. It may serve to restrict advertising so as to protect local businessmen from competition by large foreign firms. It may serve as grounds for oppressing a minority culture that can be portrayed as antinational, especially if the minority extends across the border—for example, Asians in East Africa or Kurds and Armenians in the Middle East. It can be used to combat the spread of a youth counterculture. It can be used to repress radical or participatory political ideas.

For example, it turns out that some of the support at the UN for the Soviet's position on direct broadcast satellites (see Chapter 8) came from Third World countries whose conservative governments were more concerned about checking Communists than about American propaganda. They saw the Soviet-sponsored convention as, ironically, a protection against future Soviet broadcasting.

Why American Television Succeeds Abroad

Under normal circumstances, culture needs little protection. Culture is that to which people are already attached. If the culture is satisfactory, if local media are doing their job of providing products that fit the culture, the audience will be satisfied with mostly domestic works.

The marketplace for ideas does not accept imports easily. Local products are often picked over foreign ones for various reasons:

1. *Barriers of language.* People prefer a film made in their own idiom than one with subtitles or dubbing.
2. *Barriers of social support.* Much of the enjoyment of media is in discussing them with one's friends. Reading this year's bestseller or seeing a big game is a social experience.
3. *Barriers of culture.* Allusion is a large part of what art is about. Foreign products have jokes that are harder to get, stereotypes that do not ring a bell, situations that do not come from daily life.

Despite these barriers, almost everywhere outside the Communist world, American television programs are a large portion of the schedule. Advocates of cultural integrity point to the one-sidedness of exports and imports in television programming, as in all mass communications. The developed countries (particularly the United States) export messages; the less developed countries import them.

Once a program has been produced, the costs are sunk. Whatever additional revenues can be obtained from rerunning it is net gain, so the producers will sell rights to their programs for whatever the traffic will bear. An episode of a typical TV drama in the early 1980s cost perhaps $200,000 to produce; the U.S. network probably paid the producer somewhat less than that, since both sides knew that the producer could make something from reruns. Once a producer's investment has been recouped, he has his agent sell whatever reruns he can. Iran was able to buy the program for $300–$400, Thailand for $250–$350, and Peru for $250, one or two tenths of one percent of the cost of producing it.[23] *Variety* estimates the average range of a half-hour episode of an American TV show as between a high of $4,200 for Great Britain and a low of $20 for Haiti. A poor-enough country can press the price down toward almost the cost of negotiations and mailing. The result is a large flow of programs from those countries where many programs are produced and where the producers need the revenue to countries that lack adequate facilities for production.

The largest exporter is the United States, at least in real terms. Great Britain is another large exporter, but that is partly because the United States can be charged heavily for the British programs it imports. Mexico is another large exporter because its products have affinity to the Latin American market.

Why has American programming been so successful around the globe, despite barriers of language and custom? The American television system, being predominantly commercial, is geared to producing what the public wants. And what the American public wants is not so very different from what other publics want. Cultural and language differences are important, but human universals are important, too. Americans today are able to appreciate Aristophanes, the Ramayana, the I Ching, and the Bible. So, too, the rock music that appeals to our youth appeals to youth in many countries. Batman and Donald Duck strike chords that give them their popularity here, and they strike similar sentiments with people in other countries. The American television and movie production system gives unusual priority to finding those popular chords, and this explains to a large extent why American programming has so successfully breached the barriers of cultural difference.

All producing countries compromise in various ways between the demands of the market and the agendas of civil servants or other salaried intellectuals. The philosophical differences between these two sets of selection criteria are profound. In China, producers worry about keeping the interest of their audience, though telling the people what they should be told certainly has priority. Even in the United States some production is funded by foundations and by public broadcasting, thus allowing selected professionals to do their own thing; but this plays a smaller part in the U.S. balance than elsewhere. It should be no surprise that in the country where producers try hardest to produce what the public wants, they succeed. The Americanization of world culture so often deplored might be better described as the discovery of what world cultural tastes actually are and the adoption of those into commercial media. If American pop culture is successful around the world—and it is—it is by a circular process. American commerce reflects world cultural tastes; the product in turn feeds back into the system and reinforces that which was already found popular.

Another reason for greater American success at television exports than that of some other major television producers is the market pressure on the American producers to sell syndication rights and the lack of such pressure on producers in some other countries. American independent producers depend on reruns to make any profit. Elsewhere, where a well-financed noncommercial network does its own producing, excellent programs may be produced, but there is little incentive to adapt them to the world market or, once made, to market them vigorously. NHK in Japan is the outstanding example; it has superb programs and abundant revenue from a license tax on TV sets. Soviet television is another example of a large network with little economic drive to export. The BBC, though its revenue is from a license tax and its structure is just like NHK's, does have an incentive to export because it operates financially in the red and needs the extra revenue.

Is that process cultural imperialism? If so, it is hard to know who is exploiting whom, and who is dependent on whom in this symbiotic relationship in which the U.S. television producer runs a marginal operation and depends on foreign syndication network oligopsonies. The process is far too complex to be caught in such simple ideological clichés as cultural imperialism or cultural integrity.

There certainly is cultural diffusion. There is also increasing convergence of world culture; wherever you are you will find American movies, pop groups that sound like those in Liverpool or New Orleans, artists painting like French impressionists, magazines (legal or illegal) with Marxist perspectives, and world news from one of the great wire services. Americanization is also everywhere; often it is not really Americanization but rather Westernization, for the items that come from the United States (like clothing fashions, for example) actually originate elsewhere in the West and exist all over the West, though the United States is the biggest purveyor. As the largest single capitalist country, and one of the most dynamic, as one of the largest exporters despite its deficits, and as the country with the most advanced communications industry and structure, America's impact on world culture is enormous.

Equally certain is the fact that human differences persist, that cultures remain different. Tokyo is not New York and neither city is Paris. There is a world cosmopolitan culture of which Tokyo, New York, and Paris are all parts. The sweep of that culture in a world of easy communication is irresistible; the integrity of parochial cultures is being broken as they all become urban, industrial, and part of one vast flow of knowledge and the arts. But the change is selective and integrates everywhere elements of the old with much of the new. Tokyo, New York, and Paris each remains its own place.

The Fallacies of Protection

There is little substance to justify the fear that a country's development may be hurt by the free flow of information. Every government quite properly seeks to be part of that extraordinary march of progress which modern technology has brought. In the field of communication, it appropriately seeks for its own people the facilities that will make possible the autonomous expression of their culture and the efficient development of their society.

How can a country build up its communications capability? How can a nation that now imports most of its TV programs become an effective producer of programs? How can a society with a weak press make it strong? How can a country that uses linotype, mechanical telephone switches, and handwritten records advance to on-line computer networks?

As we have seen, the most tempting panacea for upgrading a nation's capabilities is to bar foreign competition. Quotas are put on cinema and TV imports. Newspapers are compelled to use only the national press agency. Data networks are stopped at the national frontier. That kind of protectionism is a natural response to an imbalance of information flow; it is a reflex reaction to the power of foreign information processors, or to what seems to be cultural imperialism. But it is, in most cases, a misguided response; it does not do what it is intended to do; it is likely to do just the opposite.

Of course every country, or for that matter province or city, desires to improve its communications. To want that is no more than to want its people to have the ability to know what goes on about them, to express themselves, and to live better. The expansion of production capabilities to meet otherwise unmet communication needs and the fostering of indigenous culture are goals that are almost unanimously endorsed. Building studios, subsidizing the arts, establishing training programs, upgrading telecommunications, and extending communication networks to outlying villages are kinds of national policies to which there are few dissenters. What is at issue is whether these goals are helped by protectionist barriers on foreign products of the intellect. They rarely are.

It is still a debated question among economists as to when protective tariffs help those who impose them and when they hurt. But since the time of Adam Smith, any economist who defended tariffs by the simple-minded argument that the exporter must gain and the importer must lose would be laughed out of court. Respectable protectionist economists of today understand Smith's refutation of the simple version of their argument and are forced to make a much less obvious case in terms of factor costs. Any respectable argument today has to take into account Adam Smith's compelling point about comparative advantage. He demonstrated that both parties gained when each specialized in what he could do most efficiently, and then they exchanged. The same considerations enter into evaluating communications.

Politicians appeal to the prejudice that if something is of foreign origin, the receivers are hurt and the source benefits. At that unsophisticated level of discussion an American politician takes it as proof enough that Americans are hurt if he can show that Italian shoes are gaining in the market; or he takes it as a condemnation that some political idea is "un-American," just as a politician in a developing country can discredit an idea if he can show it is Yankee. Social scientists operate at a more analytic level.

Two sets of social science studies, among others, bear significantly on the process of international cultural exchange. Anthropologists, since the 1920s and 1930s, have debated the relative importance of diffusion and independent invention in the formation of cultures. Political scientists and sociologists, since the 1950s, have examined the relationship of cultural flows to national integration.

Cultural diffusion

The studies of diffusion document the following conclusions: that every culture consists largely of elements adopted from outside, but that the process of adoption is selective and adaptive. Imported elements acquire meanings and features that integrate them with the new environment and make them quite different from what they were in their place of origin. African music, for example, brought by slaves to America, became American jazz, which in turn became the music of adolescent modernizers in Eurasia who were rebelling against traditional cultures.

Of course, everyone recognizes that American culture was largely imported by migrants but in its derivativeness it is not unique; as a new society it may have a larger proportion of recent adoptions than say the culture of France, but even France is a conglomeration of foreign things. The root stocks that support the vines on which French wine grapes grow originally came from California. The towers of Neuilly and Montparnasse were invented in Chicago, and many of the cars on the street had their prototypes in Detroit. Much of the electorate vote for parties whose ideas came originally from the German Marx, or the Russian Lenin. Their religion came from the Middle East.

Ralph Linton tells it well.

> The role which diffusion has played in its growth may be brought home to us if we consider the beginning of the average man's day.
>
> Our solid American citizen awakens in a bed built on a pattern which originated in the Near East but which was modified in Northern Europe before it was transmitted to America. He throws back covers made from cotton, domesticated in India, or linen, domesticated in the Near East, or wool from sheep, also domesticated in the Near East, or silk, the use of which was discovered in China. All of these materials have been

spun and woven by processes invented in the Near East. He slips into his moccasins, invented by the Indians of the Eastern woodlands, and goes to the bathroom, whose fixtures are a mixture of European and American inventions, both of recent date. He takes off his pajamas, a garment invented in India, and washes with soap invented by the ancient Gauls. He then shaves, a masochistic rite which seems to have been derived from either Sumer or ancient Egypt.[24]

The imitation that every country engages in is selective. It copies the things that meet its needs. The selection applies to both material things and ideas. In addition to technology, developing countries want to assimilate scientific thinking, secularism (sometimes), and democracy. They integrate the ideas and objects that they adopt into structures of their own.

The choice of what to adopt and how to adapt it is not necessarily made by the government. There is no more congruence between what people want in the less developed countries and what their governments think they ought to want than there is in the developed countries; indeed, the congruence is less because in general the governments in the less developed countries are less democratic and less competent than they are where education, wealth, and participation have advanced farther.

Cohesion, division of labor, and dependency

Studies of international communication and social integration, beginning most actively in the 1950s, did not aim at as universal a theory of the process of diffusion as had the earlier anthropologists. Rather, three common theories about the impact of international communication have emerged.

The most widespread theory is that the more communication there is between peoples, the more cohesion there will be and the greater the prospect of peace. This is the theory on which people-to-people programs are based. It assumes that the better we know one another, the more we will like one another.

The same hopeful theory was expressed in the years after World War II in the formation of UNESCO and in the doctrine of free flow of intellectual intercourse. Wars, it was said, are made in the minds of men, and peace would come from contact. More realistic analysis came in the 1950s in a long series of empirical studies of the actual effects of exchange programs. Students who study abroad, it was found, do not necessarily become more favorable to their host country. There is, on the contrary, a U-shaped trend of favorableness; it declines with contact for a while before it rises again; in the end the most important changes are not to more positive attitudes but to more differentiated perceptions. So the first theory—that increased communication means increased cohesion—is but a very partial truth.

A less familiar, but older and much more sophisticated, theory asserts that the growth of international communication produces a more complex international division of labor. Adam Smith was a classic expositor of this theory. Any division of labor, such as that in a market or in a complex organization, requires that the different actors know what others in the market or organization are doing, what can be obtained from others, and what the opportunities are for one's own special activity. If communication is so essential, then expansion of the realm from which good information is available is likely to mean expansion of cooperation, competition, and division of labor.

Adam Smith viewed that as an expansion of human welfare and freedom. Karl Marx recognized the same facts. He also recognized the expansion of intercourse as making possible the expansion of social organization and of the division of labor. But he saw that as dehumanization, not as freedom. In the Marxist view, a full man is a promethean character able to do everything himself. In that view, to be confined in specialized tasks by division of labor is to be enslaved because it makes one dependent on others.

That brings us to the third theory about the relation of communication to social integration, namely, dependency theory. Dependency theory, which is particularly popular in Latin America and other Third World countries, is only a simplified derivation from Marxism. Dependency theorists argue that interactions between the underdeveloped and developed worlds make the weak nations dependent on the advanced ones. In that analysis it does not matter what the nature of the interaction is. Of course exploitation creates dependency, but, according to the dependency theorists, so does foreign aid. So does obtaining Western education, or getting news from Western news sources, or watching American TV.

Each of these three theories of communication and social integration contain an element of truth; yet all are greatly over-simplified. Growing intimacy and interaction does promote peace and understanding under most circumstances. Communication is a condition for division of labor, and division of labor does promote productivity. Also cooperation does imply mutual dependency, just as a husband and a wife are dependent on each other. And it is also true that when one cooperating partner is much stronger than the other, a dependency relation is unequal.

Yet the measure to which each of these valid but partial theories applies varies greatly with the kind of relationship in question. There are many kinds of interactions among nations. These include conquest, occupation, colonization, trade, investment, and communication. Each of these works in different ways. The feelings generated, the advantages gained, and the patterns of dependency are different.

The main variables on which these international relations differ include *equality–inequality*, the *physical presence* of one population among the other—or lack thereof—and the existence of means for *continued status maintenance*. Let me illustrate these concepts.

For example, in trade, typically, the two parties are equal at least to the degree that each may refuse a deal, though they may be quite unequal in the alternatives that they have outside the deal. The autonomy that each has and the mutual advantages are enough, however, that nations rarely want to stop trade in principle. They all say they want more trade, but of course they also want better terms.

Colonists, on the other hand—that is, agricultural settlers—typically get expelled after a while unless they take over power completely. The fact that they are physically present is an irritant. The basic inequality of their being landowners with a status protected by law is so unpalatable to the colonized that sooner or later they come to blows.

Investment is a much less exploitative relation than colonization and usually more stable, but it does require some devices for continued status maintenance. The investor wants to know that his investment is safe. Law is the usual defense of his property, though it is not unknown for such other devices of status maintenance as force, myth, and money (or bribery) to be used, too. The greatest resentment is usually created by extractive investments which eventually get bought out or expropriated. The least resent-

Table 1. Equal and unequal interactions between communities.

Means of status maintenance	Unequal		Equal
	Intruding persons present	Intruding persons not present	
Force	Conquest, occupation	Threat of conquest	Combat
Law (including property)	Class privilege	Investment	Organized international cooperation
Bargaining, exchange	Tourism	Foreign aid	Trade
Myth, tradition	Status privilege	Ideological movements	Common faith
None	Missionarizing	Propaganda	Discussion

ment is generated by those productive investments that expand opportunities for indigenous activity.

These comments are a prelude to asking about communication. What kind of relations among peoples does that kind of interaction typically produce? The answer is portrayed in part in Table 1 which compares various kinds of intercultural interaction as a function of two variables: the degree of inequality in the relationship and the devices that maintain the continuation of each party's status.

The striking point is that typically communication is unsupported by any major status-maintenance devices. The communicator may be physically present, as in the case of a missionary, or not personally present, as in the case of a foreign broadcast. There may be inequality, as in one-way broadcasting, or there may be equality, as in a discussion. But once the communicator has said something to the hearer, the information imparted has become the hearer's property as much as the speaker's. There is no way a communicator can keep control of his ideas.

Copyright is a partial exception. It is an attempt to use law to maintain the status of the originator of a statement. It is completely at the will of a sovereign state, for governments are free to write or ignore copyright laws as they wish. It is also an exception that can hardly survive in the face of new communications technologies.

So, in general, information that flows to a country from abroad is quickly appropriated by the receiving country if it wishes, and ceases being a foreign *controlled* influence. A cultural purist may regard the content as an evil foreign intrusion. But it is there to be used or not at their own will by the people or authorities of the receiving country. The imparter of the information retains no control over it.

Industry's behavior illustrates the extreme appropriability of information. Industry makes less use of patent monopoly today than half a century ago. Industrial practice today swings between the extremes of secrecy and liberal licensing. A patent is a limited-term monopoly granted in exchange for full publication of the process. But once there is publication of the process, it is usually easy to duplicate it. Organized research laboratories using scientific principles are likely to be able to find a parallel procedure. Today, also, there are many manufacturing centers abroad that are not necessarily bound by the same law. So industrial secrecy has become in many instances more valuable than a patent. Companies like IBM, AT&T, and Xerox equal the Department of Defense in protection of security. But once a product is in the market, and its secret accessible to any laboratory that examines it, then companies today find it pays to be liberal in licensing.[25] Licenses have become a major item in international trade.

There are entire computer data bases of processes offered for license. Technology transfer has become an explosive process, and many firms have found that they can't lick it so they join it.

Thus, if we look at the various types of international interaction in Table 1, we see that the degree of dependency that results varies greatly with the kind of interaction and the duration of the dependency. There are situations, like that of technology transfer, where the very fact of the interaction has a self-limiting effect on the dependency; the borrower soon appropriates the knowledge and ceases having to borrow it. There are other situations where foreign dependence acquires a self-perpetuating character. That tends to be the case in highly capitalized activities, not in intellectual activity. If an oil field is developed in country A, its customers in other countries become significantly dependent on it. If superpower B sells jet fighters to a developing country, the latter becomes significantly dependent on B for training and parts. If a foreign investor opens an industrial plant in country C, there will be a continuing relationship over many years to follow.

One must distinguish among the developmental patterns that follow on different kinds of international relationships. Examples of quite different kinds of international patterns are offered by agriculture, extractive investments, industry, and intellectual activities. Foreign investments in agriculture, whether by settlers or by creation of plantations, have profound social effects and usually end up in more or less violent intergroup conflicts because so many thousands or millions of ordinary people are directly affected in their whole way of life. Investment in industry is more favorable to development but also has a partially self-perpetuating character. The intrusion of service and intellectual activities has a quite different effect because they are labor intensive and because the initial practitioners have no way of keeping their monopoly.

The Diffusion of Centers of Activity

The fact that service and intellectual activities are so readily appropriated by the country to which they are introduced underlies a fourth and much more sophisticated theory of communication and cultural integration than the cohesion, division of labor, or dependency theories. This fourth theory, originated by Karl Deutsch, starts with the proposition that communities tend to develop where there are heavy flows of interaction, and that boundaries form where the interactions are less.[26] He tested that proposition by examining such indices as mail flows, citations in scholarly literature, travel, and trade. The conventional expectation is that as a field matures and as the world grows smaller, the proportion of material from foreign sources goes up. But Deutsch found, surprisingly, that the reverse was the case. The data revealed a regular time sequence in the balance between domestic and international flows. There is a cycle in which the proportion of international flows grows at first, and then domestic flows catch up.

Deutsch found such a pattern in the literature in physics. Early in the century there were only a few major physics centers and few major physics journals in the world. Among the centers were Germany, Zurich, Switzerland, and Cambridge, England. Everywhere else physicists read and cited the foreign sources from those few places. But as physics grew and matured from the 1930s on, departments, societies, and journals were established in various countries, and students and scholars could find what they needed in their own institutions and their own language. Scientists began increasingly to cite literature in their own country and native language, and to be trained at home rather than abroad.

So the ratio of domestic to foreign citations went up, not down. What hid awareness of this fact was that the field as a whole was growing, so international communication in physics continued to grow. There were more international meetings, more international journals, not fewer. But the domestic activities were growing even faster than the international ones. Physics in that relative sense became less international.

The same cycle can be illustrated in field after field. French impressionism, at one time, burgeoned forth from Paris and made that city the center of the art world. Emigré artists congregated from all over. By now, in virtually every country, painters work in the tradition of Matisse or Renoir, sculptors copy Rodin or Maillol. They have communities of colleagues of their own; Paris is no longer a center for them.

A similar cycle of diffusion of centers of activity is becoming apparent in broadcasting. Radio programming is very decentralized. It is produced in virtually all countries; the audio tape recorder has greatly facilitated program production in the field and towns and villages away from great centers. To some extent the late adopters of radio broadcasting have imitated the centers from which it diffused. The general pattern of programming consists of music, news, drama, and religion, plus some special programs for children, farmers, and housewives. But there is some adaptation. Some countries censor more severely than others; some stress religion, or quote the urgings of the national leader. In a few countries there is genuine innovation to meet the needs of different situations. Some developing countries, for example, have used radio broadcasting to fulfill communications functions that in more highly developed countries are fulfilled by the mails, the press, the telephone, or other media. Two-way village feedback is provided in the Senegal farm programs. In China the wired radio network doubles as a phone system.

Most of the world is still at a stage of television development in which it imports much of its programming from a relatively few production centers, most notably the United States. However, new exporting centers are developing in countries such as Mexico. In general it seems that the American lead is eroding, and indeed some day it must end. International TV flows, as a percentage of all flows, are highest at an early stage when development is very uneven.

That observation has been made even by the critics of the influence of American media abroad. The British sociologist Jeremy Tunstall notes the use of imports at an early stage: "There is no country or territory in the world whose newspapers, magazines, films, records, radio and television have not been influenced by those of the United States; and in most countries this influence has included, during the formative stages of the medium locally, the importation of substantial quantities of American-made products."[27]

A country going into television broadcasting needs, if it is to succeed, some low-cost source of second-hand programs. These permit it to fill the airwaves cheaply while its own creative organizations struggle to begin to stockpile some backlog of indigenous program material. Thus, every new television system falls back on cheap foreign programming for a while.

Gradually, as the system matures, the proportion of such imports go down. How far down they will go and how fast depends upon the size and wealth of the country and the method of funding the television system. Most observers would not wish to see imports decline to the point where television ceases to reflect the world's cosmopolitan culture. American television would be much worse than it is without imports from the BBC or the occasional European movie classic. In general, the quality of a cultural medium can remain high only by openness to using the best material, regardless of from where it comes. Nonetheless, there will always be a bias, among both audiences and producers, in favor of those things that are indigenous. Television can be counted on to use as much of that material as there is available. The problem is only how to foster the creative industry that makes such material available. Certainly one of the conditions for that is that there be a lively television system in operation. Use of foreign materials is therefore more likely to end up being a help than a hindrance to the growth of the indigenous industry.[28]

Contrary to common assertions, the new communications technologies facilitate the development of multiple production centers rather than leading to concentration. For one thing, cable TV and cassettes offer minority audiences, for the first time, a variety of specialized programming. In the second place, satellite transmission favors the creation of distributed networks. And last, the cost of cameras and editing equipment is falling drastically. Camcorders have already produced a revolution in video culture. "Video freaks" are producing imaginative material all over the world. Broadcast-quality material can be produced by equipment only moderately more elaborate than what they usually use.

Whether in developed or developing countries, these trends mean that ambitious, creative people anxious to deliver a message to interested small groups will face far less of an economic barrier in the future than they have in the past. Many of them will be talking across national borders to people of other lands who share their interests. Others will be, for the first time, producing in their own countries kinds of materials that before were obtainable only from abroad.

The growth in low-cost and simple domestic production in poor as well as in advanced countries will, however, not stop large-scale TV imports any more than the growth of physics in many lands stopped international meetings or journals. The more rapid growth of domestic production of certain types of materials will, at the same time, be accompanied in most places with expansion of the number of TV channels and growth in the total amount of TV programming. There is every reason to expect that the growth of video for entertainment, information, and education will be so large as to absorb both ever larger amounts of domestic material and also growing numbers of imports.

A small country (whether industrialized or underdeveloped) cannot produce enough programs to fill the air waves with a dozen or more hours a day of new, original TV material 365 days a year, year after year for every channel. That is just as true for Norway or Belgium as it is for Zanzibar or Honduras. The fact that it takes a market area of hundreds of millions of persons to support the enormous appetite of a TV system for new programs is indeed a severe limitation on TV's ability to reflect a single national culture. It can do that only within the less than ten giant nations with populations of 100 million or more. Elsewhere, TV production will have to be a cooperative regional or other group activity with procedures for exchange between countries. Indeed, TV will be better if all countries engage in free exchange of the best products from wherever they are made—East or West, rich or poor.

There is nothing in the economics of development or of video production that predestines which countries will succeed best in the competition to produce widely desired and successful artistic material. Filmmaking is a labor-intensive, not a capital-intensive, activity. It is expensive in the United States because artificial monopoly structures, such as enormous rewards to talent, have been built into the industry. But noncommercial filmmakers have demonstrated that high-quality film can be produced cheaply. Indeed, American commercial filmmakers are increasingly producing films in developing countries, so as to reduce costs. And India has proved that a poor country can build a major movie industry; Egypt has shown that a poor country can make a major broadcast effort; and China has demonstrated that a poor country can build a low-cost distribution system to communicate to millions.

To a large extent, the pattern of more diffused activity with cheaper equipment is what one can expect also from international computer communication. If technicians in a less developed country are trained to operate on a modern computer from a remote terminal, fairly rapidly programmers there will be trained and analysts will learn to use the computer. Soon they will be doing their own input and developing their own data bases. Very large operations may have to be done by remote computing to take advantage of facilities the less developed country does not have. But even that is a useful learning experience by which some computer technicians learn how to do things that they will use in their next generation of domestically located hardware soon after.

In short, the main barrier to entering the game is trained humans and access to a machine. Remote access provides both at an early stage. Soon minis and then larger computers will follow, once the skilled personnel exist. Telecomputing accelerates, rather than slows down, the development of computing in a country.

The same considerations apply to print media. The spread of literature from abroad is the precondition for the development of a sophisticated group of literati and a reading public. Once that comes into existence, domestically produced journals begin to appear and quickly become successful among the local audience more than anything brought in from abroad.

From the social science literature on cultural diffusion, therefore, we learn that there is an enormous amount of adoption of culture elements from abroad and that the items adopted are modified to fit the adopting culture. In this process there is a cycle of initial direct dependence on interaction with the source culture, but that is followed by a patriation of the new activity and a relative growth of domestic interactions as a result.

Status-maintenance devices

The process of patriation of foreign cultural elements is particularly rapid in those areas not under the status-maintenance devices of force, finance, or law. Status-maintenance factors in the realm of knowledge or information are varied; let us at least mention four.

Most important, undoubtedly, is governmental authority via censorship and licensing. In most countries it is illegal for a layman to hang up a shingle to give legal or medical advice. In 1559 the British government prohibited anyone from opening up a printing plant outside the City of London or the university towns of Oxford or Cambridge. In Eastern Europe it was long illegal to form an ideologically opposed political party. All these rules protect exclusivity and privilege in certain fields of knowledge. In the absence of world government, such restrictions have little or no influence internationally.

The same is true for a second status-maintenance device, copyright. While underdeveloped countries have accepted the provisions of the Berne Convention, there is nothing to compel them to do so, and some of them do not. Since the overwhelming bulk of information flow is from the developed countries to the underdeveloped ones, it is clearly in the self-interest of the developed countries to have copyright protection and collect royalties and in the interest of the underdeveloped countries not to. The fact that most countries, their self-interest notwithstanding, have signed the Convention and respect the property rights of authors and publishers in copyright is a strong testimonial to the law-abidingness of human beings. But there is no compulsion behind this.

A third factor has been very important in the past in giving some countries a monopolistic edge on knowledge but is becoming much less significant; that is the need for large capital investment in equipment used for information processing. In the first decades of the century, Great Britain had a powerful influence on international communication because of its lead in laying the very expensive undersea cables that were necessary to transmit signals across oceans reliably and clearly. Satellite circuits on the INTELSAT satellite are much less controllable and are available on ready lease to anyone. Twenty years hence when the art of satellite launching has become widespread, no doubt nothing will be left to control of international telecommunications through capital investment.

Video production, we have just seen, used to be done only in expensive studios with large camera crews; camcorders and cheap editors are changing that. Until the 1950s the printing presses that turned out books, magazines, and newspapers became more and more monstrous as they grew more efficient. A huge rotary press in a major printing plant could turn out tens of thousands of sheets an hour but would cost millions of dollars. Today, offset printing, computer typesetting, laser printers, and xerography, while they cannot compete on the biggest runs, bring small runs down to comparable levels of cost and allow print-quality work to be turned out on the office floor by secretaries. In the 1960s it looked as though giant computers in massive central information utilities might require huge investments that only rich centers could afford. Today it is clear that mini- and microcomputers and dispersed computing is the future direction of information management. Thus, capital requirements are not likely to be a centralizing factor in information management in the future.

The fourth factor that may tend to increase viscosity in the flow of information activity around the world is a natural tendency toward monopoly in certain very large data bases. As far as international flows are concerned, this is perhaps the most significant factor in preserving imbalance. Some data bases may well end up being global monopolies accessed by telecommunications from all over the world because they are offered at a price and with a liberality that makes it not worth anyone's while to duplicate them. Those countries that get a jump on data-base publishing are likely to keep it for some time. Nonetheless, data-base publishing is still in its infancy. There is still vast room for countries to get into it, and clearly many will. Also, as we mentioned earlier, some data bases are inherently regional or national in their content. Nonetheless, for this one kind of international computer communication (namely, information retrieval from data bases) our general conclusion that a lead in international computer communication is not self-perpetuating, but rather is self-eroding, should perhaps be slightly qualified.

The Theory of Comparative Advantage

The flow of culture is always uneven, but it need not be monolithic. Artists and writers congregate together and stimulate one another. They create centers such as medieval universities or modern laboratories. There are rarely more than one or two dozen great world centers in any given field, narrowly defined. But there is rarely only a single one. Usually there are a variety of centers, each with its own specialization and each representing a particular intellectual school within the given field.

Satellites and on-line computer networks support the tendency toward the creation of new and dispersed centers of intellectual activity. Initially, satellite reception and redistribution via small earth stations will increase the facility for importing computer and video software from abroad in a wide variety of specialized kinds that are not available domestically.

At first glance these new technologies may seem to create foreign intrusions, but the wiser policy is to welcome them. In general the case we would make is the classic argument of comparative advantage. Communications facilities, both hardware and software, are a resource. Any country has only a certain amount. There are only a certain number of talented writers, telephone circuits, actors and artists, available spectrum, books and libraries, computers, terminals, satellite positions, electrical engineers, and maintenance technicians. There are copious uses in demand of these resources.

Progressive national (or for that matter local or regional) policy is to expand these resources. Investment in education pays off. So does investment in microwave equipment, the building of studios, the expansion of publishing, the creation of news agencies, the extension of training, the establishment of computer networks, and the subsidization of the arts. But to try to foster these by protection from competition achieves the reverse result. Faced with scarce communication resources, sound policy is to import as cheaply as one can what is available from abroad and to use one's limited facilities for purposes that cannot be satisfied otherwise.

Protectionism in the intellectual and cultural field is even more damaging to the country that practices it than it is in the field of material trade. In both fields a country pays a significant price for not doing those things in which it has a comparative advantage while trying to do things at which it is inefficient. In physical trade, however, there may be certain self-perpetuating aspects to relying on imports. In things of the mind, however, the process of using a medium is also the process of learning to be its master. Protectionism in trade may facilitate the development of an industry; protectionism in things of the mind, however, inhibits learning.

Let us illustrate by an American example. For a long time, Hollywood did not adequately provide the American intellectual market (the 20 or 30 million of them) with sophisticated, thoughtful TV programs. It did not provide programs like *The Forsyte Saga* and *Masterpiece Theatre* for TV. The economics of the American film industry, with its union contracts, residuals, weak noncommercial stations, and so on, makes that sort of production very difficult to finance. Foreign imports have partly filled the gap. Their success has demonstrated that there is an unmet need and has led to American imitation.

Now that computer networks have become international, we see the same sort of encouragement to better domestic performance coming from them. When the ARPANET and Tymnet appeared on the European horizon, alarm spread through the European PTTs. Their monopolies could be broken. The IBM/American dominance in computing could spread to computer service activities if remote access to American service bureaus were to be more efficient and cheaper than service available in Europe. One response to such a threat of competition and loss of business might have been prohibitions and tariffs; and indeed in part it has been. But the alternative response, partly adopted, was to compete vigorously by developing their own European network, and this indeed was done. Many countries have started their own PTT-managed packet data networks. Since all conform to the X.25

CCITT standard, they are interconnected. At the same time, the European Economic Community supported Euronet, a private network for scientific and other types of research.

In culture, even more than in trade, protectionist measures are usually self-defeating. One developing country, for example, imposed a 30-fold increase in customs duties on Indian and American films, so as to stimulate its own film industry. At the same time, its authorities arrested dissident students and intellectuals and established severe censorship to protect national values. Under those circumstances, there is no likelihood of a domestic film industry growing. We may expect that the net effect of keeping out foreign films will be a shortage of films to show, the closing of theaters, and a decline in the movie-going habit; the domestic industry will suffer, too.

India, under Indira Gandhi's short-lived emergency rule, required that all news be transmitted to the press by an Indian press service in order to develop a strong national news agency. The net effect was a marked decline in the liveliness of the press. It was even more serious than the effect of censorship, for censorship covered only a few topics, but the press service monopoly covered everything. And the monopoly did not serve the goal of building a strong national news agency; that was inhibited by the decline of the press on which it is based.

Thus, competition from abroad can be useful in improving domestic communication and leading it to do better what it learns from the competitive examples. Furthermore, as we have already noted, in any narrowly defined intellectual activity there are likely to be only a few major centers in the world.[29] Not even the biggest metropolis can be a center in everything. New York, one should think, is not a center for Marxist economics, nor London for content analysis, nor Tokyo for psychoanalysis. But any self-respecting society seeks to find those areas where it has a comparative advantage. Success in that effort requires several things, but not protection from the free flow of ideas. It requires financial support; some portion of societies' scarce resources must be allocated to the particular area of effort—one has to choose. It requires talented leadership. Finally, and most important, success requires a supportive environment—a government that does not persecute the practitioners of the art, universities that promote them, and a public that is interested in what they are doing.

Such a greenhouse will exist for only a few things at any one place at a time. The world is a network of such centers. What communications technology is doing is to make the reach between them easier. That is the answer to a common complaint in political discussions at UNESCO, for example, about the one-way flow of communication. It is indeed a measure of the underdevelopment of a country if the flow of information in all things is one way. But it is not enough to show that in any one field, like news services, the flow is one way. In any segment the flow is inevitably unbalanced. The goal is not an even flow or even a more even flow; it is a more diverse flow. The goal of development is not to have a balance in every field but to be a master in one's own. The measure of leadership in the industrial nations of the world is not only how much more goes out from them by way of TV or computer networks or wire services but also how much more goes into them by way of news from elsewhere, propaganda, and publicity efforts. One does not find the *New York Times*, the *London Times* or *Asahi Shinbun* in a developing country, nor for that matter the *Financial Times* or the *Economist*. So we must set aside overly simple notions about the balance of flows and think instead in terms of specialization and comparative advantage.

In short, a prescription for raising the communications capabilities of a less developed country might be put in four points:

1. Specialize initially in some activities where the country has the greatest potential for becoming a major center of activity and focus resources on those activities.
2. Import as much and as cheaply as possible the vast stock of knowledge and cultural materials that a country needs to develop both audiences and skilled professionals.
3. Reject rules, such as copyright, which established sources try to impose to prevent the appropriation of knowledge. Knowledge flows freely and is essentially unmonopolizable, but those who own it sometimes try to introduce friction into the process. In a world of sovereign nations, the disadvantaged ones need not accept such restrictions.
4. Establish a permissive and supportive milieu for creative people; let them read what they want, write what they want, travel, and mix freely.

There are very few countries following this prescription. Under the circumstances, it is hard to be persuaded that their failure to develop a communications capability is because of a flow from elsewhere. Yet the lesson will not be heeded. Much of mankind will penalize itself by timidly trying to defend itself from the benefits of new communications technologies. And yet, in the end one can make an optimistic prognosis.

Despite the highly protectionist and restrictive attitudes that prevail in both the United States and other countries, the potential of useful applications of telecommunications technology is likely to be realized. The restrictions are likely to be only delaying factors. The reasons for making that prediction is that the world is a competitive place. Protectionism protects inefficiency, removes incentives to self-improvement, penalizes consumers, and lowers the gross national product. These points have become part of the accepted philosophy in North America and Western Europe in the last couple of decades; the same awareness is likely to arise in the field of communications, at least in some places. Those countries that are restrictive of innovation (and they will probably be the majority) will lose out in competition to those countries that take full advantage of the new possibilities.

Chapter 8

Broadcasting from Satellites to Home Receivers: A Case Study

On November 9, 1972, by 102 to 1, the United Nations General Assembly referred a Soviet draft convention on television broadcasting from satellites to its Committee on the Peaceful Uses of Outer Space for consideration. The minority of one was the United States. The convention would have outlawed direct broadcasting into any country without its consent; the vote was only a procedural one to refer the draft to a committee for study. The issue itself was unimportant, but it symbolized something very important: the way in which new revolutionary technologies of communication have alarmed most established authorities and led them to try to use the organs of the international community (the UN, UNESCO, and the ITU in particular) to place controls on the free flow of information.

In this chapter I shall examine the early debate on direct broadcast satellites, first recounting the technical facts and then noting how the different nations in the three main groups (communist, developing, and industrialized) voted, the arguments they used, what was really on their minds, what technical assumptions they made, and the consequences of what happened. The same could be done for five other issues concerning the free flow of information that have been debated in international forums:

1. the Third World News Pool
2. assumption of responsibility by states for the international communications stemming from them
3. prohibitions on disapproved propaganda
4. short-wave broadcasting
5. spectral and orbital allocations

I note the significance of each of these issues at various points, but it would be tedious to narrate the details of the debate on each. I shall use the direct broadcast satellite debate as a case study: the story in each of the other cases is much the same.

In 1972 the shock of finding itself in total isolation in UN votes was new to the American delegation. The trend had started some time earlier, but this was a whistle that signaled to many who had not yet noticed it that the game had changed. It signaled, too, how complete had been the swing from 1948, when the Universal Declaration of Human Rights had been unanimously passed recognizing the right of people to "receive, and impart information and ideas through any media and regardless of frontiers," to the situation 24 years later, when an overwhelming majority of states regarded transmission of television across frontiers as a proper matter for concern and control.

By the 1970s a new majority of states thought less of individual rights than of the wishes of governments. Also in 1972 UNESCO, an organization originally created to promote free cultural exchange, adopted a Declaration of Guiding Principles on the Use of Space Broadcasting which contained such injunctions as:

II. 2. Satellite broadcasts shall be essentially apolitical.
VI. 2. Every country has the right to decide on the content of the educational programmes broadcast by satellite to its people.
IX. 1. It is necessary that States . . . reach or promote prior agreements concerning direct satellite broadcasting to the population of countries other than the country of origin . . .
2. With respect to commercial advertising, its transmission shall be subject to specific agreement between the originating and receiving countries.

Those principles passed 55 to 7, with the United States in the minority.

That same year the Soviet Union, not satisfied with an unenforceable UNESCO declaration, introduced its draft convention on direct satellite broadcasting at the UN. Many years later, the debate continues.

What was it that so many governments were afraid of, if foreign states could broadcast TV to their people? There were numerous pontifications throughout the UN debates about the enormous power of television to manipulate a society. The Canadian–Swedish working paper of 1969 quotes the UN Space Conference as saying: "No single development can perhaps alter the face of the developing countries as television via satellite can" (p. 9). It adds, "In developing areas with undeveloped communication systems, broadcasting from satellites could dramatically change the entire outlook of millions of people" (p. 55). "The introduction of television programming, based on practices, from one social context to another and greatly different one, might result in a cross-cultural shock whose consequences are not yet fully predictable" (p. 74).

The summary statement of the 1969 U.S. House of Representatives hearings referred to broadcast satellites: "In the hands of a totalitarian state, this instrument might be used to impose the kind of social control imagined by George Orwell in his nightmarish novel, '1984'. Or it could engender propaganda 'wars'." The delegate from Pakistan, Ali, said: "The vividness and immediacy of television sets it apart from other information media in terms of its impact and influence over the minds of the recipients. The experiments performed some years ago in the field of subliminal suggestion indicate the possibilities for abuse or misuse to which a strong medium of communication like television can be put" (p. 72).

What was feared was the undermining of entire countries by political propaganda, pornography, and TV violence. Various delegates expressed their worries in the discussion:

Baroody (Saudi Arabia):

What if the United States, or any other State for that matter which in the future develops satellites for international television broadcasting, refuse to adhere—leaving aside participating—to a convention? We who do not possess the technology of outer space . . . would have no guarantee . . . that we would not be the victims of salacious programmes that might undermine our culture, our way of life, our traditions (p. 81).

In the free enterprise countries the motivation of newspaper men or broadcasting companies, privately owned, is gain and profit; so they pander to the cheap—or perhaps "sensual" is the word rather than "cheap." They pander to the sensual and give people the things they like to read about, peddling sex (p. 37).

Ake (Ivory Coast):

[Direct satellite broadcasts can] easily be used for the . . . purposes . . . of subversion and propaganda, which might harm both the sovereignty of States and internal public order (p. 28).

Grigorov (Bulgaria):

The appearance of such stuff is another kind of aggression, without bombs or bullets but with consequences at least as serious for the security and welfare of the peoples. . . . Television programmes aimed at other States could openly praise war and aggression, justify colonialism, present in a favourable light abhorrent systems such as *aparthied*, advocate racial hatred and so on. We remember the radio broadcasts of Hitler's Germany, or another and even more blatant example in this field: the propaganda carried on by television in favour of *apartheid* and racialism in South Africa (pp. 43–45).

Malik (USSR):

The representative of Belgium spoke of the free flow of information. But a question arises, whose flow? A clean flow, a creative flow in the interests of peace and mankind? Or is it to be polluted by sex, violence, propaganda, misinformation, slander, interference in international affairs, against the culture and civilization of every single nation? (p. 46).

Pohl (El Salvador):

It is not by any means a question of trying to regiment art or to set up yardsticks . . .; rather it is a question of . . . avoiding bad taste, vulgarity, the commercial exploitation of sex and other excesses which characterize not liberty but licence (p. 16).

Such fears persist, and the arguments about the dangers of direct satellite television continue to be heard, despite the increased sophistication of technical knowledge.

The story of the early direct satellite broadcast controversy is almost an allegory about how difficult it is for political bodies to understand technical complexity; it deserves some detailed accounting. To a large extent the debate has been about imaginary technological dangers unrelated to reality. To laymen or science fiction enthusiasts, satellite broadcasting was a prospect without limit. From the time of the first Sputnik satellite, writers have foreshadowed the day when TV pictures beamed from space could bring education and cultural exchange into TV sets in every village, desert, island, and city of the world. "With the help of a large Sputnik," said Soviet Professor S. Katayev in December 1957, "Moscow television programs could easily be relayed not alone to any point in the Soviet Union, but also far beyond its borders."[1] Arthur Clarke, in an oft-quoted remark, talked of the technical means to "drag the whole planet out of ignorance." The U.S. information effort, too, sought every occasion to magnify the achievements of the American space effort. It became recognized that if Americans could put a man on the moon they could put a picture on a TV set anywhere on earth.

Since the scientifically unsophisticated UN delegates were in no position to question assertions that direct broadcasting from satellites was technically easy, they had to accept assertions that it was just around the corner. The usual projection in the 1960s was that it would come within ten years. However, the technical realities were distinctly more complex, as we will see in the sections that follow.

Direct versus Redistribution Broadcasting

The first and most important distinction is between broadcasting from a satellite *directly* to the viewer's television set versus *redistribution*, that is, transmission from the satellite to a ground station, which then retransmits to the home set.

There is actually a finer graduation. At one end is the present system of low-power redistribution satellites which transmit to large dish antennas in ground stations costing dozens of thousands of dollars. At the other extreme is direct high-power broadcasting to home receivers. This requires a more sophisticated, more powerful satellite, whose signals can be picked up by a small home antenna costing a few hundred dollars at most. There is a tradeoff between the cost and power of the space segment and the cost and power of the earth segment.

Between the two extremes lie any number of intermediate possibilities, of which perhaps the most interesting is the so-called satellite community or master antenna. For example, 12-foot dishes made of chicken wire were put into villages by the Indian SITE program; at $2,000 apiece they were too expensive for an individual home but affordable if shared by an entire village. Higher quality antennas often span 15 feet across and are used by cablecasters. These may also be considered community antennas, for in a country like the United States, there could be tens of thousands of those, serving as redistribution nodes for every town and even neighborhood if needed.

The choice between direct broadcasting from satellite to the home versus a distribution from a satellite to terrestrial transmitters for conventional rebroadcasting or cablecasting to homes is essentially an economic one. When Vikram Sarabhai first designed the Indian SITE experiment, he had to decide whether it made economic sense to transmit separately from the satellite to the single TV set in each village or whether transmissions could be picked up by ordinary sets in each village, without benefit of the $2,000 chicken-wire antenna. The answer came out for a hybrid system. In the Ganges plain where villages were close to one another it was cheaper to attach the antenna to a relay station so many villages could be served at once without an antenna in each. Where villages were more scattered, the better solution was to put a dish in each village.

That sort of trade-off calculation has to be made for each situation. But for some situations the factors are so one-sided that there can be little doubt about the outcome. With any technology that we can now foresee (that is, at least for the next two to four decades), TV broadcasting directly to each home will not be the normal method of TV distribution in urbanized societies. There are other situations where the set owners are scattered far apart and where there has as yet been no investment in microwave relays and relay transmitters to reach them. People live in Eskimo villages, in the wide spaces of Siberia, on remote islands, in the Amazon jungle, and in the deserts of the Middle East. Granted these are small populations, but they are important ones to reach. About a fifth of the Soviet population lives in remote areas where TV reception is poor or lacking. About two percent of the population of Japan live on remote islands or deep valleys and cannot receive broadcast TV. About 50,000 people inhabit the Canadian North. It may well be a matter of public policy to get TV broadcasts to them even at substantial cost. Indeed, TV may be socially far more important for them than for an urban population that has many alternative forms of culture.

From the point of view of international communications, direct broadcasting has the advantage and disadvantage of not going through a local relay. Sovereign nations worried about foreign broadcasters sending unwanted programs to their people do not have to be concerned about redistribution satellites, for the authorities can censor the material at will at the relay station. Internationalists seeking to promote contact among peoples may regard that as a disadvantage of redistribution systems. That is why the debate in the UN has been about direct broadcast satellites. Yet direct satellite broadcasting is unlikely to become a particularly important part of the overall world broadcasting picture, as we will see.

Frequencies and Orbital Locations

The layman's image of direct broadcasting from satellites is conditioned by his experience with radio; he chooses a radio station by a flip of the tuner; directionality of the antenna is generally at most a secondary adjustment to improve reception. For various reasons the situation in regard to satellite reception is not the same. That is partly because most of the usable spectrum was already assigned for other uses when satellites came on the scene, and partly because of engineering reasons.

When satellites arrived in the 1960s the question was where to find spectrum for civilian communications with them. Fairly large blocks of spectrum were needed. The standard satellites of the 1960s and 1970s (like INTELSAT 1, 3, and 4) had twelve transponders, each with enough bandwidth to carry a television picture. Their aggregate bandwidth was comparable to the entire allocation for VHF television with its 12 channels. Newer satellites had 24 transponders, and the number of satellites was increasing rapidly. Where was spectrum in such amounts to be found? The answer was mostly in the previously large, unused area above 11 gigahertz.

For technical reasons, the frequencies chosen had to be in the microwave range, in which the beam can be focused. A telecommunications satellite is functionally a microwave repeater hung up in the sky. A beam pointed at it from earth is received by the satellite's antenna and then relayed (at a different frequency) back down to earth. Any part of the beam that goes off into space or to the earth without hitting an intended antenna is wasted energy. So the more narrowly focusable and aimable the beam the better for point-to-point communications; in that way one can put more of the radiated energy to telecommunications use.

Even within the range of microwave frequencies and above, not all frequencies are equally desirable. Depending on their purpose, engineers have strong preferences. For a reliable low-cost telecommunications service, frequencies in the standard microwave range (like GHz) are desirable. Developed equipment and techniques exist for it, and larger dishes are needed than for the frequencies above 11 GHz. But that was not a major disadvantage from the point of view of a telephone company. Telecommunications planners were still thinking of using only a few earth stations as intercontinental links between established networks. Cheap ground stations were not a major consideration, but rather highly reliable long-lived satellites.

For direct broadcasting the requirements were quite different. Low-cost antennas are essential. They must therefore be small; thus, high frequencies not shared with terrestrial uses are desirable. On the other hand, in frequencies above about 12 GHz the beam acquires more and more of the characteristics of light, including attenuation by rain.

All of these issues came to a head for the first time in the World Administrative Radio Conference (WARC) of 1971. Until then the ITU had been a club of engineers solving technical problems, unwatched and uncared for by politicians. But in the decade of the 70s the politicization of international organizations spread to the ITU. Block voting and attempts to label some countries as pariahs began to occur at ITU conferences, though most political objectives were camouflaged in engineering jargon about frequencies, flux densities, or types of modulation.

In 1971 Argentina moved an explicitly political resolution on the content of programs. That was withdrawn. Underneath the discussion lay the fear of direct TV broadcasts beamed by one country into another. One means to prevent that was by allocating spectrum for satellite use only in ranges and with low enough power to preclude direct broadcasting to existing TV sets, or by allocating spectrum for shared use by satellites and terrestrial functions, which would restrict direct broadcasting because of interference. All proposals from the United States and other countries for satellite spectrum assignments below UHF were solidly defeated by countries concerned about interference and direct foreign broadcasts.

UHF band proposals were made by the USSR, Canada, and India but were opposed vigorously by the U.K. and other European nations concerned with interference. The final compromise was that "while assignments could be made in the 620–790 MHz band using FM, the low level of the power flux-density agreed upon made direct-to-the-home broadcasting unlikely."[2]

In the higher frequencies where there is less congestion and no home television, there was less controversy. In the 2.5 GHz band an allocation was made but with severe limitations on power flux density. The most frequently proposed band for satellite use was 12 GHz. A total of 14 national administrations proposed allocations in this band. The negotiations were similar to those at 2.5 GHz. There was no outright opposition to admitting the service in principle, but widespread disagreement on the status of the assignments, owing to markedly divergent needs and priorities. The United States and Latin America clashed over the issue of the status of satellite service vis-à-vis terrestrial services in Region 2 (the Western Hemisphere). The United States pushed for priority to the satellite broadcast service while an iron-clad Latin American coalition pushed for their equal status with the terrestrial allocations. A highly complex set of allocations resulted from all the infighting. Allocations differed by region, as did elaborate footnoting within each region. Region 2, for example, provides two satellite-type services on priority status but are limited to domestic systems and are subject to previous agreement among the countries concerned—a compromise between the United States and Latin America. There were a wide variety of proposals in the range above 40 GHz but little controversy. Thus the first and only exclusive, worldwide allocations for broadcast satellites appeared.[3]

In addition to allocating spectrum for satellite services at frequencies such that direct broadcasting would not work with existing TV sets, the 1971 WARC adopted further restrictions on transmissions from one country to another. These restrictions were phrased not in terms of broadcasting but in terms of radiation from one country that might interfere with radio activities in another—which is, of course, the central purpose of the ITU. Resolution No. Spa 2-1 adopted at the 1971 WARC requires prior coordination between countries if the frequency that one of them proposes to use is thought likely to have harmful effects on radio communication services in a receiving country. There is well-established precedent in Article 9A of the Radio Regulations for such consultation whenever a country is concerned about possible interference with its own services. But in addition to that, a clause was adopted which many commentators have argued represents an explicit prohibition on satellite broadcasting to any country without its consent. The U.S. government does not agree with that interpretation; but, however interpreted, Clause 428A is a very severe limitation. It says: "In devising the characteristics of a space station in the Broadcasting-Satellite Service, all technical means available shall be used to reduce, to the maximum extent practicable the radiation over the territory of other countries unless an agreement has been previously reached with such countries." In the view of the United States, this is not an agreement to allow censorship of programs, something not within the purview of the ITU, but only concerns consent to levels of radiation on given wavelengths. The line is a fine one; one cannot transmit content without radiation.

The anxiety of most countries that direct international television broadcasting might come into existence, just as direct shortwave radio broadcasting had, was not put to rest by these restrictive spectrum allocations and rules against interfering radiation from abroad. Further restrictions kept being proposed. Much of the debate concerned the desirability of having an international plan for satellite broadcasting. That there should be such a plan was agreed to in the first week of the conference in a resolution promoted by France. Pending such a plan, which would have to be worked out at a future conference, the United States wished to develop satellite usage under present ITU rules; the USSR wanted an embargo on all activities until such a plan was adopted. Spa 2-3 and 428A were the compromise.

The design of a plan was the subject of the 1977 WARC. That conference dealt with one specific frequency, the 11–12 GHz band. The importance of that frequency was that it had come to be seen as the natural location for any direct broadcasting from satellites. Higher frequencies were on the edge of the state of the art and suffer from rain attenuation. Lower frequencies are congested, and the nations of the world were not motivated to upset existing allocations and broadcasting systems at great expense to provide something that they were quite ambivalent about, namely, direct satellite broadcasting. The 11–12 GHz band was therefore the most promising allocation for direct broadcasting.

A major issue concerned the assignment to specific countries of specific orbital positions for future satellites and also specific frequency channels for broadcasting from those positions. The United States opposed the adoption of a detailed plan as premature. As a result, no plan was adopted for what the ITU calls Region 2, the Western Hemisphere, and an early freezing of decisions was avoided for the Americas.

For the rest of the world a plan was adopted. The less developed countries were particularly anxious to adopt a system that would give them each an equal piece of the orbital resource. The radio frequency system had developed in a different way: first-come, first-served. The WARCs would allocate large slices of the spectrum to particular services, such as maritime, aviation, broadcasting, and so on. When a country wanted to use a frequency for one of these purposes, it would look within the appropriate band for a hitherto unused frequency and then register a claim for it with the International Frequency Registration Board. Thus the countries that moved first gained a squatter's right over those that came later. From a technical point of view that is a good system, since the best frequencies get used first. However, less developed countries view that system as a way in which the rich countries get theirs first; the resource may be gone when the less developed countries are ready to use it.

There is, of course, a simple solution to that problem. Each country could be given rights in the spectrum and in the geosynchronous orbit according to a plan, but with the right to sell or rent them. In that case the rich and the first entrants would simply pay the poor for the privilege of using, for the time being, that which the "spacelord" does not himself then have a use for. That solution, however, offends the ideology of much of the world. The managers of spectrum are impeded by the notion that it is somehow improper to create a market in what is a social resource of all mankind's. It has somehow seemed less offensive to moral sensibilities to try to resolve such conflicts of interest by political infighting in conferences and by the adoption of rigid and complicated international agreements. The moral difference is not obvious, and the benefits to poor countries are clearly greater with a market.

In any case, the obvious market solution was not seriously considered, and as a result of WARC 1977 an elaborate plan for the Eastern Hemisphere was adopted. It gave each country (down to San Marino) at least five channels for direct satellite broadcasting. In Region 1 (Europe and Africa) a total of 40 frequency channels were assigned,[4] and in Region 3 (Asia and Oceania) a total of 24, all exclusively for broadcasting. In Region 2 (the Americas) the band was also assigned to fixed and terrestrial services. To avoid interference, the broadcast satellites for the United States, Canada, and Mexico must be located between 75 degrees and 95 degrees W longitude (those for South America 75 degrees to 100 degrees) or 140 degrees to 170 degrees W longitude; that allows separation of the beams from those of the fixed services. A fuller plan for Region 2 was adopted at a regional meeting in 1982. In the Eastern Hemisphere the plan assigned frequencies and the specific location on the orbit for each country's broadcast satellite. Special arrangements were made for some regional groupings, such as a joint Nordic satellite system.

From all of this it should be apparent how great the difficulties are for one country to broadcast directly by satellite to another country. First, an available and appropriate orbital position relative to someone else must be secured. In some cases that would be easy; neighbors could be reached from one's own orbital positions, for example. But it would be much harder for the United States, for example, to broadcast to the Urals on the other side of the world.

Second, the broadcasting country would have to find a frequency in that range which TV sets in the receiving country can tune to and which is not in use there, or indeed will not even cause interference with the frequencies in use there. After all, if the broadcasting is to be direct, the signal has to be a strong one. On a given frequency a weak signal for local retransmission might not cause interference, but a signal strong enough for home reception most likely will.

More often than not, the conditions for transmitting a direct broadcast to a particular foreign country—namely, a well-placed orbital position and an available frequency—may not exist, but what will almost always exist is some reception of TV across the transmitting country's frontiers. Neighboring countries, like neighboring cities, have had to work things out so that if A uses channel 2 near the border, B does not use that frequency. There is no feasible way to prevent some Canadians from being in the range of U.S. television, nor some Israelis in the range of Jordanian, nor some Estonians in the range of Finnish. Similar border spillover will occur with satellite broadcasting, too, and potentially over a much larger area. Borders do not follow the orderly ellipses of radiation footprints.

Cooperative versus Unwanted Broadcasts

Consider first cooperative broadcasting. The receiving country has to arrange for distribution through its markets to its population of a large number of low-cost earth stations. Normally it will do so only if it is engaged in direct broadcasting, too. Since satellite transmissions are highly directional, the dishes have to be pointed quite accurately at the transmitting satellite. Normally they will have been installed pointing at the country's own satellite. To enable receivers to point to more than one or two satellites, a device for accurate shifting of the dish would have to be added. In a dictatorship, citizens pointing their dish to foreign satellites would be at some risk, given the visibility of the antennas.

A large part of the negotiation about the 1977 ITU plan was to find groups of countries that were happy to share the same orbital position; such clusters increase the difficulty of broadcasting to countries with a different ideological position.

Finally, the receiving country will have to leave a frequency available for the foreign transmitter within that range of frequencies for which the earth station electronics is adapted.

Cooperation of the type just described may often occur, for example, in Scandinavia. But the question that upsets the UN delegates is what happens when one country tries to force propaganda broadcasts on another whose government objects. Suppose, for example, that the Soviet Union distributes a significant number of small earth stations in Northern Siberia and the Mongolian desert and launches a powerful satellite over the South Pacific to broadcast to them. Suppose the United States then launches another satellite only a degree or so away. Under those circumstances, if the Soviets were to use only some of the channels on their TV receivers but had wired their antennas to receive all the available channels and convert them, then, of course, the U.S. broadcasts on that free channel could be seen, though not quite as well as pictures from the correctly pointed Soviet satellite. But that is a scenario that requires Soviet cooperation. Under any normal circumstances, the invading broadcaster would have no real prospect against a space-power government that was determined to prevent invasion.[5]

Developing countries are not primarily concerned about a scenario of superpower conflict. They are more worried that somewhere where local television is not well developed a foreign country or company will put out satellite broadcasts and sell television sets with dish antennas to receive them. However, there are few poor countries whose control of foreign exchange is so loose that they could not either prevent the sale of antennas or take advantage of it by insisting that the vendor let the country broadcast things they wished to transmit on his satellite.

A foreign country that wished to broadcast uninvited to the few direct receivers that someday will exist would need to launch a large, expensive satellite capable of beaming high-power radiation on a band not otherwise in use in that area but suitable for satellite transmission and receivable by sets currently in use. If it wished to reach more than one country, it might need separate satellites for various longitudes, with different standards of signals to different countries. These requirements would rarely be met on any significant scale. To invest in direct satellite television broadcasting to countries that do not cooperate would require great expenditure for minor results. The situation is entirely different from short-wave broadcasting, which takes little bandwidth, can be broadcast with relatively cheap transmitters, and uses simple receivers that many people own anyhow for other purposes than to listen to the propagandist. Direct satellite TV broadcasting to countries that do not wish it is not an attractive option for either governments or commercial broadcasters. It is not something rational people should be worrying about.

Most satellite broadcasting today and in the future will be redistribution broadcasting, not direct broadcasting. Some of it, particularly outside of urbanized areas, may be directly received, and some of that will be done multinationally. But the issues that actually result from such multinational activities will be small compared with the very serious problem of spillover to neighbors who are not always friends and the problem of television exports. These are real issues in the international flow of television that engender strong feelings. There is every reason to expect continued frustrations and tensions on issues surrounding the flow of television, but the reality of direct satellite broadcasting will be a minor item among those concerns.

The American, Soviet, and Third-World Positions

One reason for the intensity of early debate about direct satellite broadcasting is that in international organizations there is a kind of Gresham's Law: phoney issues drive out real ones. Real issues are sometimes too hot to handle because the good guys and the bad guys tend to be all mixed up. For example, on the question of television and cultural imperialism, the real issue is the importation of American and other programming by underdeveloped countries. But that is not a good issue for UN debate because the programs that were imported were bought voluntarily by the poor countries. The United States is at worst a co-conspirator, not the lone villain. On the other hand, direct broadcast satellites beamed to foreign peoples are an ideal issue. No one actually loses by anything that is said or done about them, at least immediately.

Though the issue is phoney, it is not unimportant. Precedents are being set. Resolutions and conventions are being proposed and debated for the imposition of regulation on the flow of international communication. From a tactical point of view those countries that would like to stop the free flow of communication have chosen their issue well, because the defenders of free flow, most notably the United States, are not inclined to be intransigent and waste their credit on an essentially secondary issue. And so the American reaction has been repeated compromise. Gradually, regulatory precedents were set, such as clause 428A. At a World Administrative Radio Conference in 1985, several planning principles were adopted to further the resolution adopted in the 1979 WARC, to "guarantee equitable access in practice" for all countries to the geosynchronous orbit. In 1988 the second session of the 1985 WARC convened to complete the Geosynchronous Orbital

Allotment Plan. Each country was allotted a space on the geosynchronous orbit, effectively making the establishment of a satellite in another nation's slot illegal.

The restriction of the free flow of information is a real issue, not a phoney one. The problems created by free flow do not come just from bad guys, and the answers to the problems are not all simple. The problem of avoiding spectral interference, for example, requires regulatory procedures. This and other concerns of the less developed countries are very real, but the pat answers that have been offered are wrong. One reason why that is not apparent to those involved is the lack of technical sophistication of those engaged in the political debate. On the direct broadcast satellite issue it was NASA representatives from the United States who had the facts. But as an interested party, they were not readily believed. For one thing, in this age of scientific marvels, laymen find technologists particularly noncredible when they denigrate their own capability. When the magic-maker says he cannot perform as desired, he is suspected of protecting a covert interest. The scientist who says pollution levels cannot be brought down far enough is suspected of serving industrial profiteers. In the satellite broadcast debate a widespread reaction to those who minimize the prospects has been suspicious disbelief. The argument begins "if you can put a man on the moon, then . . ." And to detailed explanations of technical problems, the layman's reply is that these problems may vanish with scientific breakthroughs.

The American position on direct satellite broadcasting is predicated upon the belief that the free flow of information across borders is a good, not an evil. Thus the American strategy has been to make satellites, outer space, and information flows via satellites accessible to all on the freest possible terms. NASA launch capabilities, EOSAT resource mapping capabilities, and INTELSAT communication capabilities are available to all.

The traditional Soviet position is based on the premise of prior consent: No country should be broadcast into unless it agrees to it in advance. The theoretical basis for requiring such consent is the doctrine of sovereignty. The American worry is that if the doctrine of sovereignty is accepted in the direct satellite broadcasting context, the notion that sovereignty requires prior consent to broadcasts would apply equally to short-wave radio.

An additional point to be noted is that acceptance by the United

States of the principle of prior consent by other countries to broadcast into their territory might well be unconstitutional. The American government would, if it accepted that rule, be bound to prevent persons in the United States from transmitting messages via satellite that other countries chose not to accept. The U.S. government would be in the position not only of enforcing our own laws against libel, sedition, unfair balance, and copyright violation—that is, laws that conform to the First Amendment—but also the laws and policies of the receiving country. That is not a posture that the United States government can easily assume.

The Soviet delegation and some others have understood that the prior consent rule, which they advocate, is incompatible with the American practice of leaving the content of what is broadcast in the private domain, not subject to control by the government. To avoid such devolution of responsibility they proposed to affix responsibility on governments for whatever is broadcast. That too is clearly incompatible with the U.S. conception of freedom of the press and was rejected by U.S. representatives in the UN and UNESCO.

What the developing countries fear is cultural imperialism. Broadcast satellites, they often believe, will impose an American commercial and alien culture upon them. But there is much to be gained by the poor nations from a free flow of information between nations. Satellites used for earth observation, seastate measurement, meteorology, shipping and aviation control, international computer networking, and international communication have much to contribute to productivity and development. Satellite intelligence has helped stabilize deterrence by reducing the prospects of military surprise.

Faced with the *fait accompli* of satellite information gathering and satellite communication free of all political restrictions and controls, and faced by American administration of this new technology in a fairly generous and liberal way, the nations of the world welcomed some of the benefits to be gained. They recognized in the International Convention on the Peaceful Uses of Outer Space that the playing of tactical games with one another's satellites and with satellite communications would only deprive themselves and the world of valuable assets.

So far the nations of the world have behaved well in outer space. They have not interfered with one another's satellite activities. They have ruled that sovereignty does not extend to outer space. They have cooperated technically in using satellites. They have generally accepted that satellite weather and earth observation information should be in the public domain and that INTELSAT should be open to all nations that wished to join.

In the near future there are prospects for considerable further advantages from satellite communication for the poor nations. As we shall see in the next chapter, small, cheap local ground stations and the development of satellites appropriate to them can provide communication to and from villages far sooner and more cheaply than could be done by way of terrestrial microwave relays or cables.

But if space so far has provided a remarkable example of progress, liberalism, and international cooperation, it is clear from the satellite direct broadcast story that xenophobic suspicion, protectionism, and striving for sovereignty has been hiding close by in the wings. The political authorities are more than ready when threatened by any loss of control and an opportunity to do anything about it to rush into the breach with a system of regulation and control.

Chapter 9

Communications for the Less Developed Countries

By the standards of the year 2050 we are all developing nations. There are rich nations and poor ones, powerful and weak, but the treadmill moves fast. Once Argentina enjoyed a standard of living equal to many European countries. Two decades ago Lebanon was a modern and prospering country. On the other hand, less than fifty years ago Germany and Japan were the destroyed losers of a world war, and Israel, Singapore, Taiwan, and South Korea were poor and underdeveloped. Compound growth or compound decline makes its mark.

It is not true that the rich nations all grow richer and the poor nations all grow poorer; it is true that the gap between those that grow and those that do not grow becomes ever more cavernous; and there are many countries that have chosen to put pride before growth.

Labor skills and motivation, natural resources, energy supplies, peace, and an inflow of capital may each be more important than communication in explaining the economic success stories of Japan, the United States, Taiwan, Germany, and South Korea. But communication is clearly a factor in development. As a country becomes an information society, communications become increasingly important to its growth. In this chapter we will try to understand how communications policies contribute to development or to its strangulation.

The word "development" is ambiguous. The economic denotation of the word is the clearest one; it means increasing GNP, productivity, and living standards. In a second sense development means a multiplication of centers of initiative. "Development" in that sense borrows from Darwinian biology, systems theory, or the theory of organization. A simple organism, be it a living thing or a society, takes direction from one or a few centers; it can make limited responses determined by those centers; it has a simple division of labor between the centers of command and the rest of the mass. A developed system is one with a complex division of labor, where initiative is taken in many places and where the results can be forecast not from any one center's will but only from complex interactions in the system.

In still a third sense development means something that is better referred to as "modernization" or "Westernization"; but people avoid that usage, fearing the charge of ethnocentrism or cultural imperialism if they suggest such directionality of the process. In that sense a developed society is one with a participant polity; it is an open society with status determined by achievement rather than ascription; in it, relations are universalistic rather than particular; it is secular, literate, and urbanized.

Here, my use of the word "development" is most often in the first and second senses. No doubt the growth of international communication and mass media have been the main hypodermics for the injection everywhere of ideas of democracy, equality, revolution, science, pop culture, sexual liberation, secularism, consumerism, and change. The global diffusion of a Western style of life is patent for all to see, though so is the counter-reaction of adaptation of what is diffused and reassertion of parochial identities. These points about the content of development today are important, but they are not what we are discussing here. Rather, we are focusing on the relationship between the development of modern communications and the multiplication of centers of initiative in society—and with that a growth in productivity. Low-cost long-distance communication, as we shall see, is increasingly producing division of labor, multiplying the centers of initiative, and thereby raising economic productivity. Modern attitudes and values we shall be concerned with here only to the extent that they motivate economic growth and productivity.

Communication is having an impact on productivity in both rich lands and poor lands. The theory of development is thought of too often to be a theory of a special case: the less developed countries. It is assumed that the economic and political processes in them are different from those in the "normal" developed countries. The very use of the word "developed" for rich lands reveals an absurdity in this train of thought. The grammatical form implies a process that has ended; "developed" countries, if the grammar were taken seriously, have gotten there and have nowhere further to go. Clearly that is nonsense. Normatively, development is particularly important for those countries that are poor, but veridically the concept applies equally to those that are rich. Granted, development in the less developed countries is a goal worth caring more about than further affluence in already-affluent societies, but most of the same analytic statements apply to both.

If there are differences between the process of development in rich lands and poor lands, industrialized ones and peasant ones, it is because, in a process of separate but combined development, those countries that join the game later learn from those that joined earlier. The latecomers can skip stages that the pioneers went through step by step. Those who come late can acquire in their full fruition technologies that evolved elsewhere by trial and error. Those who came early have sunk large investments in plant that is obsolete. Germany's and Japan's industrial success since 1945 has arisen in part from the very destruction of their old plant. The use of direct satellite broadcasting and electronic switching may come in the future more easily in Iran or Indonesia or China than in the United States or Europe, because there is no existing system to be abandoned.

It is a continual game of leap-frog. Countries that latch on to a technology first may find themselves by-passed and regain the lead only if they leap once more. Telephone dialing, for example, began in Europe after 1900,[1] whereas in the United States among the independents it came a decade later, and in the Bell System almost two decades later. The big Bell System had invested heavily in manually switched plant and was reluctant to use scarce capital to make the shift.[2]

While the transition to automatic switching was seriously delayed in the United States by the financial problems of large-scale conversion, competition was ultimately a stimulus to action. In 1912 the Automatic Electric Company claimed 300,000 phones on automatic exchanges.[3] The Home Telephone company, a major competitor, had a dial patent; in 1910 AT&T absorbed it. After 1920 the Bell System began a complete shift to dial phones.

Technology transfer is one reason why the gap between the poor countries and the rich countries on the average narrows.[4] Contrary to the popular cliché that bemoans the widening of the gap, the statistics show the opposite. Countries short of capital and other resources can actually grow somewhat faster than the advanced industrialized countries, partly because of the advantage of coming second and being able to borrow rather than invent advanced technology.

Against that advantage is the terrible disadvantage that these nations lack capital and skills. Each dollar of production turned from consumption to investment is a far greater sacrifice from an income of $300 per capita than it is from $3,000. In that fact lies the moral imperative for the transfer of capital and expertise from the rich countries to the poor ones.

Do Poor Countries Need State-of-the-Art Technology?

What use are satellites, computers, television, and telephones to countries where loads are still carried on porters' heads and water is drawn at wells? Are they a luxury for a small elite in the capital cities and for the convenience of foreign businessmen, or do they contribute to the needs of the people? Is investment in communication a priority use for a poor country's scarce resources?[5]

The crucial issues were raised by Daniel Lerner in 1958 in *The Passing of Traditional Society*.[6] Lerner's hypothesis was that the growth of mass media has a psychologically broadening effect which *causes* modernization and development. Others sought to challenge Lerner's assertion of a causal relation, arguing that the undoubted presence of a correlation between development and the growth of communication could be explained just as well by the opposite direction of causality: do the mass media cause development, or do they just come with development? To test which way the causality goes, social scientists tried statistical causal models. The Lerner hypothesis stood up well.[7] It is, of course, not a sole cause, but the evidence is strong for there being a causal relation between the diffusion of mass media and modernization.

The possible mechanisms through which this causality may take place are many. Lerner based his thesis to a large extent on the psychological process of broadening of empathy. Others have noted the breaking of the cake of tradition, the diffusion of innovations, the raising of expectations, and the broadening of the arena of political and economic action.[8]

The press, movies, and especially radio operate in many ways to foster development. They convey knowledge of new ways of doing things. They raise aspirations. They create identifications on a national and even international scale. They help create a wider market for goods and a less provincial political arena.

Yet few underdeveloped countries have taken full advantage of the powerful instrument of development available to them in the communication media. Most fear mass media as a Pandora's box they cannot control. They fear the revolution of rising expectations that may come from watching video pictures of people riding in cars or eating in fancy restaurants. They fear the consumerism that people may learn from ads for canned goods or refrigerators. They fear the loss of traditional values if people read the news in *Time* magazine, or the words of Chairman Mao. And so fearing a genie that they may not be able to control, most governments of developing countries have not devoted resources to communication comparable to what they invest in transportation, education, or health.

Today, new means of communication are becoming available that make it easier to bring mass media to villages. Satellites can deliver television to rural Indians and Eskimos. They also provide telephone communication and establish consultation between the village nurse and the doctor at the district hospital. Villagers can tune their transistor radios to get the national news reports, or market reports from the local town, or world news or propaganda from abroad. They can even be put on line to a data retrieval system.

The information that the developing countries need in order to progress comes in a multitude of ways. Students go abroad for education. Colonists and missionaries in the past, and technical assistance personnel and businessmen now, settle for years or even decades at a time and introduce new ideas and practices. Trading minorities such as Lebanese or Chinese move in and around, and so do refugees. Foreign mass media messages arrive in books, movies, international broadcasts, and wire service reports. These are rediffused by the domestic media. There are government information programs and educational efforts.

Some of the communication is commercially motivated, either to sell products or the media themselves. Some of it is ideologically motivated, produced and disseminated by political movements or religious groups. Some of it is humanistically based, as with educators and foundations. Still other communication is generated and passed along just for fun and recreation. All forms of diffusion contain a mixture of what is useful and what is not.

Technical assistance programs

Among the various kinds of information, of particular importance is technical information that professional persons look up in reference sources as they work at productive and scientific operations. The size of the archives needed for such purposes is growing rapidly. The less developed countries of the world require large injections of information in the form of technology transfer if they are to grow and develop. The means for acquiring it, however, have been largely beyond their reach. Today, a chasm separates the research facilities available in developed countries—such as the Library of Congress, the New York Public Library, the British Museum, the Moscow State Library—from the extremely limited research facilities in most of the world. There is no conceivable way in which a newly developing nation can create conventional libraries of that sort for itself in less than half a century.

With the development of data-base publishing to cope with the flood of new information, the position of the poor nations will worsen. They will not have access to those new information resources unless some method to facilitate information transfer is devised. Given their lack of money and skills, they cannot be expected to make much use of modern information retrieval systems or computer communication at home; however, if low-cost global data networks become available, they will be able by remote access to use the best information sources that exist. If, as seems likely, international data communications networks using packet-switching technology make access to data bases available from anywhere in the world at communications costs less than that of the mails at present, the information gap could rapidly be narrowed. A researcher in a university or planning office in a country without adequate reference sources of its own could retrieve a fact from whatever data base he wished anywhere in the world for a communication charge little more than the cost of a domestic telephone call, something he could afford.

In short, as advanced countries increasingly transfer their reference materials from hard-copy libraries to computerized retrieval systems, the underdeveloped countries will either begin to catch up in information capacity or will fall further behind, depending upon whether they are linked to these new information stores by telecommunications or not.

Along with lack of capital, one of the main things that poor lands lack is technological and scientific expertise. Filling that gap is a long, slow process. It is inconceivable that every developing country will build for itself a Bell Labs or a French CNET. Indeed, even most developed countries are too small to create all these resources for themselves. International telecommunications would allow many large-scale intellectual research facilities to become resources for the whole world. If underdeveloped countries or impoverished regions of dual societies could put themselves on-line by telecommunications to the information and expertise they need, they could accelerate their progress.

One example: a major problem in technical assistance programs is the traumatic cessation of interaction between experts and trainees that comes once physical contact is broken. Trainees from a developing country are sent abroad for months or years. When they go back home, the principles learned do not quite fit the new situations, but the teachers with whom they had had daily interaction are too far away to consult. How much better it would be if the foreign training were followed by a period of frequent telephone consultation with teachers as the trainees tried to apply what had been taught. Indeed, not only the trainee but also the trainer would learn much from such interactions.

The same trauma of termination also occurs when experts are sent into the field in another country and then return home. Silence may descend unless the experts were able to clone themselves in the brief time available. Equipment operators in the foreign country have no one to address questions to when problems arise. Thus, within a few months or a couple of years the equipment on which the expert was sent to advise may stand idle or the methods learned may be abandoned. To avoid that, experts often are kept around too long; they stay in the field when all they are needed for is an occasional comment; the work is already being done by people they have trained. How much better it would be if the technical assistant could leave yet remain in contact with co-workers in the field through a phone call from time to time.

Sometimes technical assistance consists not so much of people as equipment. A piece of machinery is in a plant in a remote location. Perhaps there was some training done and technicians sent along when it first came in. Some years later it breaks down. Typically, it may stop for days or weeks on end, while a technician is flown in or a part is sent for repair. How much more efficient it would be if the local maintenance technician could be in direct communication with the place where the machine was made as she tries this or that procedure to repair it.[9]

One common reaction to such suggestions for sustained interaction of advisors and trainees via telecommunication is fear of dependency. It is a misplaced fear. Far from becoming dependent, practitioners in less advanced places, if given better access to advanced knowledge, will more quickly acquire the skills to become independent. Suppose, for example, there was a system for providing medical advice to doctors, nurses, and paramedics by online query to data bases and if necessary to live specialists. Suppose in each country this was connected by a satellite circuit to a global system. Undoubtedly, at first the National Library of Medicine at Bethesda, Maryland, would be one of the dominant sources. But doctors and medical practitioners everywhere would quickly improve their competence. Soon specialized sources of information would begin to grow up in the most unexpected places, perhaps in China on acupuncture or in India on some tropical diseases. What might start out as a star network would soon become a dispersed network with many centers.

Or let us suppose that a global computer network were to give economic planning organizations everywhere access to the econometric models that exist today only on a few computers, such as at the University of Pennsylvania. Initially, much of the use would come from less developed countries which wished to have access to computer models in a few advanced centers. But economic statistics by their very nature are decentralized in origin, and so is expertise in them. Japanese economists would have as much interest in the information that the Indonesian planning agency generates on the Indonesian economy as Indonesian economists would have in Japanese economic plans. As people in each place learned to program and work with the available languages and models, the flow would become many-sided. And the countries that started out with most to learn could benefit the most.

Suppose that a less developed country put its own computer

center with its presumably limited facilities on line on a large data network and as a matter of policy encouraged statisticians and programmers to use the best facilities available, not necessarily the local ones. Gradually, but fairly rapidly, young programmers would learn new skills and so would analysts. Soon they would be doing things that they could otherwise not have done and developing expertise that they otherwise would not have had. They will start making their own software and developing their own data bases. The main barrier to entering the game is trained humans and access to a machine. Remote access provides both rapidly. Telecomputing accelerates, rather than inhibits, the development of computing in a less developed country.

One element of the computing situation that is less flexible, less a function of personnel training, and more a function of fixed capital investment is the computer itself. In the past, that was a massive capital item that less developed countries could not easily support. But in this day of powerful mini- and microcomputers, the cost of the hardware is not the major bottleneck to acquiring effective computer power.

Many authorities in developing countries do not accept this sanguine view of their ability to acquire independent capacities in the process of joint activities. They do not share the Deutschian thesis of a cycle of patriation of foreign learning. That defeatism is fundamentally misguided. Knowledge strengthens those who receive it. Creating barriers to its dissemination will, in the end, leave the developing countries that follow a restrictive policy weaker and more dependent, not stronger.

Intermediate versus high technology

Another objection often raised to developing countries' investing in modern communications is that they should be using intermediate rather than high technology. They must learn to walk before they learn to run. Poor countries, it is said, have unemployed labor and very little capital; they should use labor-intensive processes, like hand-set newspapers, not capital-intensive ones like computer composition. Besides that, it is argued, they should use things that they can make themselves (like the Indian chicken-wire SITE antennas), not things that they have to import. Also, it is said—and this is more to the point—there exist in every society traditional forms of communication, like folksingers, plays, or sermons, that have a credibility and meaningfulness in the culture which no imported impersonal form can have.[10] All of those arguments and particularly the last have a great deal of merit, but they do not lead to the cliché so often leaped to, that high technology communications are inappropriate to developing countries; that is a nonsequitur. Only a few counter-examples will show how shallow that cliché is. Suppose that at the time that the transistor radio first came on the scene someone had said that developing countries should not use that new technology but should stick with conventional tube radios. Yet transistor radios are cheaper, more reliable, easier to produce indigenously, more portable, and less wasteful of electrical current. From all points of view they were what developing countries needed. And as a result they have spread everywhere. Radio has become the developing world's most pervasive medium. In Asia and Latin America where modernization came at an earlier date than in Africa, the establishment of newspapers played a major role in the instigation of change and in the growth of national political movements. Then in the 1950s and 1960s radio became the dominant medium of mass communication there. Africa, where the hold of colonialism broke only in the 1950s and 1960s, never experienced the growth of an important press; it went straight to a radio-based communication system. Thus, even the process of straight imitation does not lead to a duplication of earlier sequences in development by those who come later.

It would be a mistake for poor countries to simply imitate the media institutions and practices of the wealthier ones in other ways, too. In the United States and Europe, television was introduced as a household good; in India, it is properly being introduced as a village facility. In the West, broadcasting is done almost entirely over the air; in China, much of it is done via wired loudspeakers, because wired loudspeakers do not incur the monthly cost for batteries. The communications needs of the less developed countries are different from those of the developed ones and will continue to differ in the future.

It is often assumed that if a less developed country cannot make its ordinary telephone system work satisfactorily, then it certainly cannot use sophisticated new technologies such as a packet-switched computer communication network using satellite transmission. But this is fallacious. There are some advances in technology that make operation and maintenance easier as well as some that make it harder. One has to examine the particularities of any advanced technology before one can assess its appropriateness to a pre-industrial environment.

For a telecommunications system to serve an underdeveloped area effectively as a bearer of technical knowledge, it must have certain characteristics. It should be able to be an adjunct to expression by those persons who have credibility in the culture. It should use the language and the symbols of the culture. Its contents should be capable of local adaptation. It must be cheap. It should require as little foreign exchange as possible. It must also be reliable and relatively rugged, and not require highly sophisticated maintenance and operating personnel. It must operate even in the absence of an elaborate infrastructure of stable electric current, nationwide microwave relays or cable networks, and smoothly functioning telephone service. Finally, it must link the underdeveloped region at its will to any possible sources of data, not just to ones in a favored metropolis on which it is dependent.

Those are not unachievable requirements. There are high technologies, intermediate technologies, and primitive technologies that meet many of these requirements, and there are high, intermediate, and primitive technologies that fail. Anthropologists know, for example, that the high technology of a Polaroid camera is better than a standard camera because it does not depend on a commercial infrastructure for developing and printing. On the other hand, if one has the talent, drawing with pencil and paper does the same. In general, one of the characteristics of the new communications technologies is return to more natural modes of communication. Many of the new inventions are designed to overcome the artificiality in use or burdensome cost of media that in the nineteenth and early twentieth centuries achieved wonderful things by sacrificing some measures of naturalness, spontaneity, interactivity, or ease of use.

So in the range of communications devices that are most useful in underdeveloped areas are some very primitive means like a folk singer with a guitar, a tatzpao (the Chinese handwritten wall poster), or a traditional opera; some simple technological devices like a mimeograph machine, a slide projector, a wall newspaper, a super-8 movie camera, or a wired loudspeaker; and some very sophisticated devices like electronic telephone switching, satellites, and computer data networks. We do not have to make a case for the value of the simple devices; their usefulness is recognized. What is sometimes denied is how well some of the new technologies match the requirements of poor lands.

A Four-Media Communications System

A four-media communications system seems likely to be appropriate in poor countries with large terrain and scattered villages. The four media are radio, satellite TV, satellite telephone, and computer store-and-forward message delivery.

The use of transistor radio is familiar. It is the means by which news spreads and by which the government can reach the people.

Television adds something to the effectiveness of educational broadcasting and to the quality of village life. However, sets are too expensive to be owned by many individuals, and television's appropriate use in a poor country is as a community facility. Where territories are small and densely populated, terrestrial transmission to the village is economical, but where the territory is large and dispersed, the appropriate means of dissemination is by satellite, as pioneered by Canada with its programs to the Arctic and by India in the SITE program.

Satellites also permit the dissemination of telephones in villages throughout the nation without waiting for the spread of the terrestrial cable or microwave network. Where countries are too small in size and population, video cassettes rotated around a region can also substitute for over-the-air broadcasting.

The communications satellites now generally available were optimized for the needs of the developed countries, not for the needs of the less developed countries. The current commercial satellites are all what we call conventional. These all radiate a rather weak signal so as not to create interference. Reception of that weak signal requires a fairly large and expensive dish. A large INTELSAT dish is 95 feet in diameter and commercially available, and so a ground station can cost several million dollars. At considerable loss in efficiency one can operate with smaller antennas. Fortunately, the signal strength of satellite transmission is becoming greater, and thus satellite dishes can move to much smaller size.

Such dishes can be locally produced. In Indonesia the Bandung

Institute of Technology's Dr. Iskandar Alisyabanha advanced the use of electronic technology for the development of his country through indigenous production and the use of satellites for village communication. The laboratories at BIT designed an earth station dish that can be manufactured in Indonesia and also a "village teletype." The last is an extremely simple mechanical device that can be manufactured at low cost and repaired by an average village mechanic. With equipment such as that at the remote ends of the network, international computer communication is well within the state of the art for the less developed countries.

Less developed countries can also benefit by attaching to their administrative telephone systems terminals with store-and-forward capabilities, because they can thereby increase the utilization of limited circuits. Store-and-forward message switching is a technology particularly well suited to the need of the less developed countries because it bypasses, rather than depends upon, the problems of the ordinary phone system. For an ordinary voice telephone call, both parties have to be on the spot at the same time, which can be very frustrating if it takes a couple of hours to complete the circuit. With a store-and-forward system the message sits in the transmitting computer until the circuit is available and then gets sent to the terminal for which it is destined. That may happen at any hour of the day or night. Furthermore, low-cost satellite receivers can be installed in remote regions, bypassing nonexistent or overloaded microwave or coaxial long lines. Moreover, sophisticated data processing operations can be done by remote access to locations where the prerequisites exist. All that the local service must provide is the capability of carrying low-speed code such as that which currently delivers telex or telegrams. The main limitation on local facilities at the grass roots is the servicing of terminals, which suggests that in many locations simple CRTs and simplified rugged typewriter terminals will be the order of the day for a long time. These will convey orders to more sophisticated equipment that is located where it can be easily serviced.

An important aspect of these four media is that they are not easily confined by the boundaries of nations. Unlike cable and telephone networks, and unlike airline routes which fan out from the old colonial metropole to its former colonies and dependencies, the radio and satellite systems described above allow communication from any point to any other point, freeing users to seek advice or information wherever they wish.

"Development Communication" versus Infrastructure

Thirty years ago, when the discussion of the relationship of communications to development began, no one who wrote about the subject paid attention, except in the broadest sense, to the specific content of what the media were saying. The media of communication were seen as an infrastructure for a modern society. Daniel Lerner, for example, contrasted traditional people who learned what was going on from conversation in a village with those whose source was local radio and city life, and with those who used foreign broadcasts and reading; he compared the resulting breadth of empathy in each group and confidence in their own knowledge.[11] Media that portrayed ways of life that the listener or reader had not personally experienced, Lerner argued, broadened their user's perception of the world and so modernized attitudes. Lerner did not bother to ask what specific messages an audience had received; to a considerable degree he treated the medium as the message.

That too had been the approach of earlier literature on the impact of the press. For example, in the vast literature on the press and politics in the nineteenth and early twentieth centuries, writers have not explored the specific messages in newspapers in order to clarify what the rise of a press meant for the rise of democracy. The important point the authors have made about the press was that as an institution it helped mobilize political parties.[12] Papers were founded by movements to promote their views, and journalism provided a living for professional partisans. The subscribers received something that regularly reinforced their identifications, and true believers were told what to think. Later, entrepreneurs discovered that the press could make money. A commercial press was born. By avoiding sectarianism and by playing to the ordinary man's tastes, these emerging press lords built circulation and advertising. They produced papers designed to please a wide segment of the population, not just followers of one party. Thus, in the twentieth century commercial papers drove out party papers to a large extent. The press, which at one point had been a main stimulus to democratic political participation and to partisanship, often became the opposite. As a provider of mass entertainment in place of group organization, and as a vehicle of unaligned criticism, the press came to be viewed by all politicians, regardless of party, as an adversary, not as their instrument. Note that this analysis, insofar as it is true, traces the impact of the medium itself as an infrastructure, and does so without reference to the details of what it might say.

That was the way in which the subject of communication and development was approached thirty years ago, when the conclusion was first propounded that the spread of the mass media is one of the main causes of development. The conclusion was both challenged and confirmed in the same structural terms.

But then a new and normative thesis began to be aired. Activists with moral programs began to note that while the media might bring change that could be defined as development, it was not the change or development that was desired. The mass media might induct millions into new and freer ways of life, including the assimilation of pop culture, but they might not do anything to improve agricultural practices, to upgrade public health, to strengthen national identities, or to curb the baby boom. Should such normatively ambiguous changes be called development?

A large research literature emerged which, if one were to take its conclusions seriously and literally, would seem to prove that the media were powerful when serving as agents of evil but somehow impotent when on the side of good. An odd alliance of ad salesmen and their critics both argued, each for their own purpose, that commercial advertising was very powerful in selling people whatever the advertiser chose to sell.[13] The literature on sex, violence, and propaganda likewise stressed the power of the media. On the other hand, every study of education, or of public service campaigns, or of efforts in developing countries to persuade people to adopt new and better practices came up with essentially pessimistic conclusions about the effects.[14] It could hardly be that communication is powerful whenever one disapproves the results and impotent whenever one approves them; nature is not that attentive to the observer's preferences.

Whatever the explanation of this strange perception, the perception itself led to a purposive and directive approach to the study of communication on the part of development activists. Unlike some more cynical analysts, development activists did not see communications as powerless; there were many undesired effects of communications that they perceived as great. But on the other hand they were frustrated by their own failures. So the question was raised, how should communication efforts be conducted so as to better serve positive purposes of development? What has to be done to get farmers to adopt better practices or couples to have fewer children? Out of this development communications approach came a number of interesting findings about which techniques work and which do not.

A regular sequence is found in the course of adoption of innovations, whether by peasants or housewives or physicians.[15] Exposure to the idea may often come through a mass medium, but a decision to try it requires reinforcement by face-to-face opinion leaders.[16] Radio broadcasts to Indian peasants, unless supported by organization, might increase knowledge but they rarely produced action. In villages where farm programs were heard by unorganized peasants at home, the farmers listened and learned but did not try what the broadcast advised. In villages where the program was heard in organized listening groups, the group would discuss what they heard and would often decide to act.

Findings of that kind were incorporated into what came to be called a theory of development communication. At UNESCO meetings and similar forums the point was often made that it is not enough to simply flood the developing countries with mass media; what they required, it was said, are development-oriented communications that would meet their special needs. The mass media, unguided, would merely swamp the culture with commercialism, imported entertainment, and Western-oriented news and propaganda—irrelevant at best and more likely destructive, it was argued. What the developing countries need is not Mickey Mouse and *Dallas* but information that would help achieve their exigent national goals in production, nutrition, population control, literacy, and national identity. Information resources in poor countries, the theory went, are too scarce to be squandered on the pleasures of the middle class; they must be used for higher priority goals. A related theme, as we have seen, held that cultural integrity must be protected and that the flood of mass media amounts to cultural imperialism, which in turn generates dependency (see Chapter 7).

So there are two quite different theories of the role of communication in the process of development. Both of them recognize its importance, but only one proposes that the very existence of communication is part of the infrastructure for development; the other proposes that certain kinds of content will teach people what they need to know to progress, while other kinds of content inhibit progress. These two descriptive theories, in turn, lend themselves to use in justifying two quite different normative approaches in development. The infrastructure approach is populist; it defines development as the dispersion of autonomy and initiative throughout the society, and so it sees communications facilities as tools to be put in people's hands for them to use. The "development communication" approach is tutorial or elitist; it starts with a conviction among educated idealists as to what the masses in preindustrialized countries must learn, and it evaluates communications by whether they are teaching that or not.

While total consistency is, fortunately, rarely found among people of action, it may be helpful to spell out ideal types of each of those approaches.

The infrastructure approach

In its ideal type, an advocate of the infrastructure approach will argue that development is above all a mass phenomenon. It is a dispersion throughout the society of independent judgment, self-confidence, and initiative. Development is therefore a psychological process; people in a developed country are characterized by more breadth of perception, future orientation, receptiveness to change, and motivation to achieve. Development is the creation of a new citizen. It is not development, in this view, if peasants use a better fertilizer or a better strain of seed just because they have been ordered to; it is development if they become experimental scientific farmers, regardless of what fertilizer or seed each happens to choose. Development must stem from many centers, not just from the government. Indeed, no government, no party, nor any single authority is wise enough to decide what needs to be done in all the parts of the complex system of a society. Wisdom is widely dispersed, and people on the spot are likely to make better judgments about the problems they live with than any planner in the capital. When a centralized authority preaches at people as to the details of what they ought to do in their own familiar activities, the result is alienation and distrust of the leadership, not progress.

It follows that communication facilities are a resource that needs to be put in the hands of all of the millions of people who are trying to do things in a society. The communication requirements of any society are manifold. The government requires facilities to mobilize its population; the population in turn must have the means to express itself. Business needs communications for management and marketing. Security forces depend upon communications, and so do educators, railroads, and airlines.

Communication resources need to be localized. Every provincial governor, district administrator, village chief, factory manager, rural educator, local health worker, union organizer, salesman, and innovator needs communication facilities. They should exist everywhere; local publications and radio stations should be encouraged; so should advertising. And communication needs to be two-way. It is not enough to provide loudspeakers and radios to instruct the population; it is essential to provide forums in which the experience and desires of the people can be expressed.

The "development communications" approach

In its ideal type, the opposite view, which emphasizes development communications, starts with an essentially pessimistic evaluation of the state of affairs in an underdeveloped country. The great mass of the people are bound in tradition, uneducated, and hard to persuade to improve their eating habits, limit their families, give up dowries, or change age-old prejudices and sexism. What is more, the resources for change are scarce and should not be squandered. Fertilizer, water, steel, and machinery are scarce, so their use must be planned to give preference to priority activities and not drawn off for the use of the affluent few. Communications resources likewise are scarce and also need to be reserved for priority use, in this view. Television or radio cannot, as in a rich land, be allowed to be used for fun or for special interests. The media have a development mission and must be guided by the central government to carry that out. What is more, the social fabric in poor countries is fragile. The danger of secession, of military coups, of riots, and of unprincipled exploitation is great. There is no tradition of civic-mindedness or responsible democracy, nor is there much critical judgment in the population. It is therefore essential, in this view, to keep the flow of communication under tight control to assure that the messages are constructive and not subversive.

The evidence of social research and of success in development is overwhelmingly toward the infrastructure view.[17] Yet no government would act as if it had no responsibility for influencing the choices that are made. Any sensible policy advisor would urge the government of a developing country to use its scarce media resources to encourage those attitudes and practices essential to modernization. Civic pride, inculcation of national goals, discouragement of drug abuse and racism, propagation of good health practices, and equality for women are certainly things any rational modernizing government will try to encourage. There is, therefore, a measure of validity to both ideal types.

The dilemma is not an unusual one; it is the standard paradox of democratic reform. What should a reformer in power do who both believes he knows what is right and at the same time does not believe the holder of power should or can impose his views? In developing countries with a traditionalistic populace, with prejudiced castes and tribes, with an inequitable heritage of rights and privileges, the dilemma for the modernizing leader who understands that real change has to come from the roots is how far to inflict his view on a society that is not ready for it, or how far the process has to be allowed to work its ambivalent way through. Certainly no rational leader will deny himself the right to use his power to some degree to carry out a program of advice and education based on the insights about what works that have been learned from the research on development communication.

On the whole, nevertheless, the development communication approach must be admitted to be a failure. Apparent successes with this approach are not due to persuasion alone but occur in situations where persuasion merely reinforces strong motives of economic or security advantage; successful information campaigns are ones that tell people how they can get something they already want. And that, contrary to common belief, is precisely what advertisers do. Few commercial advertisers waste money trying to reform public tastes outright. They do not try to sell ice boxes to Eskimos. The first step in any advertising campaign is market research to find out what the public wants. Only then does the seller turn to advertising to explain how his product can fit those desires. That is why advertising seems to succeed. Reform campaigns, in developing countries or in advanced ones, that set out to infuse new values, new goals, and a new way of life into the public assume a much harder task, one that cannot be achieved by mass media alone. Only religious movements have succeeded in producing that kind of change in people's lives.[18] Rulers in developing countries who turn in frustration to their media specialists and demand that they sell national goals more effectively, as if that were simply a technical job, are fooling themselves.

SITE

The Indian SITE experiment in direct television broadcasting to villages illustrates the problem of how to best use communication technology to influence individual choice. The program was conceived by Vikram Sarabhai, a physicist who headed the Indian space and atomic energy programs. The main motive was to get information into Indian villages in order to make them more tolerable places to live. It was Sarabhai's unique persuasive power and conviction, and the widespread respect for him, that won commitment to this project from Prime Minister Indira Gandhi and from the American government, which had to provide the experimental ATS-6 satellite for the test year of broadcasting.

Sarabhai understood that in the political atmosphere of the late 1960s and early 1970s in India, a project with such heavy U.S. involvement could easily have been shot down by its many anti-Americanist opponents (including the Indian broadcasting establishment), and so he kept the programming effort totally Indian and sedulously independent. That meant, particularly after his premature death, that it was in the hands of a small group of idealistic Indian intellectuals. They were convinced that broadcasting to the villages should not be a replica of the more conventional entertainment-oriented urban television broadcasting that All India Radio was simultaneously developing. Broadcasting to the villages had, in their view, a mission: to teach and to elevate in a situation where the other instruments for disseminating literacy, scientific knowledge, better health practices, and agricultural improvements were few indeed. Television was being tested as to whether it could achieve a quantum jump over what was being accomplished by half-trained school teachers, All India Radio, and occasional and inadequate village-level workers.

The tutorial instincts of the programmers were reinforced by an unanticipated development: the declaration of a state of emergency by Prime Minister Indira Gandhi. Ironically, the project, which had been seen as a reinforcement of Indian democracy, came on stream in 1975–6 in the brief period of dictatorial rule in India. In that period the government directed all the mass media to sell the specific tasks that the Prime Minister set for the country. Indeed, the government instructed radio stations to submit a log for analysis in which, opposite each broadcast program, was entered which point in her 20-point platform that broadcast had promoted. In preparing software for SITE, four goals were specified: increased adoption of family planning, improved health and nutrition practices, increased adoption of recommended farm practices, and "political awareness." A large amount of rather well-done programming in support of those goals was prepared in time for the experiment. Widespread skepticism about whether the Indian production teams would have the software ready by the time the satellite was moved over the Indian Ocean in 1975 for a year of broadcasting was confounded by an excellent on-time performance.

From its beginnings, another aspect of the SITE program was inclusion of well-designed and rigorous evaluation research. There were studies done under the auspices of NASA and parallel studies done independently under the auspices of the Planning Commission; that those studies confirm each other gives credibility to results about which people would inevitably be skeptical if they came only from the self-interested SITE organization. There were studies that used anthropological participant observers, studies that kept counts on the audience, and interview studies. There were studies of the effects on children who used the in-school daytime programs and studies of the general population exposed to the evening program. The interview studies were of rigorous design. The sample included a set of villages that got the TV programs and a control set of similar villages that did not. In both sets of villages there were interviews before the SITE program began, in the middle, and at the end. As a result of all this research, we know pretty well what effects the SITE program had and what effects it did not have.[19]

What it did not do was change the behavior of villagers in the ways that the government had specified. Adoption of new practices in family planning, nutrition, health, or agriculture was not measurably higher in the villages exposed to television than in the control villages. That should have come as no surprise. Numerous studies showed that mass media preaching, unsupported by local organization, does not persuade people to adopt unfamiliar practices.

But television—even in one year, and even with only one receiver in each village and only about an hour of programming each evening—did have important and measurable effects. The effects were perhaps more important than if they had been mere obedience to government injunctions, for the effects were in people's own ability to think about problems. In general they have reaffirmed what Lerner found in the early days of radio in the developing world. In the villages with television there was an increase in people's willingness and ability to answer serious questions. There was an increase in the knowledge that they could bring to bear on a problem. For example, language learning speeded up among children. Among adults, the answers given about questions like family planning became more detailed and specific. People were learning about family planning from the broadcasts. The electorate of India turned Mrs. Gandhi out of office a few months later, to a large extent because of the excesses of the government's family planning activities. The fact that in the 2,500 villages receiving television people had learned more of what the government was saying on the subject did not mean that they suspended their power of criticism and independent judgment any more than did the rest of the Indian electorate. The viewers understood the broadcasts and learned from them; they did not necessarily do what the broadcasts told them to do.

Viewed as an infrastructure for learning, for judgment, and for participation, the village television program in India was a success. The existence of the sets made the villages that had them a different kind of place in which to live, and the people who watched and lived in that changed environment were more able to be citizens of a modern society. Viewed in a tutorial way as a device for selling a particular program, the SITE effort may have been a failure; but the failure is rather that of a misguided notion of what communication is for and what development is all about than of the broadcast program.

The time has come to swing the pendulum back to the infrastructure approach of the 1950s and early 1960s. The expectation that communications will somehow sustain the particular program that a given reformer favors is unrealistic. That is not a denigration of the importance or value of communications. On the contrary, what communications do for a society is something far more important than aiding special perceptions of truth. Communications facilities are instruments by which all elements in a society can be enabled to participate in influencing what the society should be. Their contribution to development is not so much to promote a particular program as it is to diffuse to everyone the means whereby they can be effective in achieving their own goals.

Organizing at the Grass Roots

If communications facilities are to be provided as an infrastructure for participation by the masses in the life of a society, then it follows that the places at which communications may originate must be numerous. In many developing countries the policy of the central government is to keep the sources of mass communication strictly confined to the center. In many of them there is a genuine danger of fission, tribalism, and war-lordism. A broadcasting station or newspaper can be a powerful instrument in the hands of a dissident group; the station is one of the first sites that a rebellious group tries to seize. During the Vietnamese war, for example, the South Vietnam government strongly resisted American suggestions that regional and provincial broadcasting stations should be established. Their fear that a station might be seized by the enemy was understandable. The price they paid for security was that their own provincial officials lacked a valuable means to establish authority in their provinces. They could not talk to their villagers and thereby become familiar and respected. At the same time, a similar situation existed in Tanzania. There was a single radio broadcasting station in the capital city of Dar-Es-Salam. It was surrounded by a high hurricane fence with barbed wire on the top and was tightly guarded by reliable troops. With some difficulty it reached the whole country, as well as South Africa and other places abroad. But there were no provincial or local radio stations; they were suspect as sources of tribalism or dissidence.

Implicit in that policy is a misperception of the bases of national unity. The head of state, far off in the nation's capital, may be venerated by provincial peasants or other ordinary people if there is already a basic identification, but he is too far from their life experiences to create by himself such an identification. The band of villagers is formed mainly by their relations with people they know, such as local administrators, tax collectors, policemen, landlords, and priests. A positive national identity is not created by by-passing those important intermediaries, however much they may be possible sources of dissension, but only by a chain in which they are the intermediary links.

This issue has been much discussed in the history of political theory.[20] The classic statement of the centralist point of view was formulated in Jean Jacques Rousseau's *Social Contract*. To Rousseau, all intermediary structures between the individual and the general will in which the citizen submerges himself are destructive of freedom; he condemns estates, parties, and representative parliaments. That is the part of Rousseau's theory that has been seized on avidly by totalitarian theorists ever since. Marx's detestation of division of labor, the Latin American opposition to freedom of association, the fascist and Soviet notions of the expression of the people's will through the single party or leader that embodies it all have an intellectual history that goes back to Rousseau.

There are, of course, plenty of examples in developing countries of courageous efforts to distribute means of communication down to the grass roots. There are many wise rulers who recognize that their effective national leadership depends upon providing village and province chiefs with ways to become known to their people and thereby to acquire personal local influence. Many governments realize that the national solution to problems of health or agriculture consists of the solution to many local problems of health and agriculture. There are also many who understand that the solutions are not the same in every place and that local judgments must go into local solutions.

An interesting statistical study by Charles Murray in Thailand bears upon this.[21] He examined a sample of solutions of local problems that were repetitive and discrete enough so that one could score in each case how successful the solution had been and count the distribution of successes and failures. Matters like roads and water supply are examples. In each case he characterized the outcome not only in terms of success but also in terms of whether initiative had come from top or bottom, and what the relationship was between local and provincial authorities and experts. He found, as one might expect, that the type of problem made a difference. There are problems for which external expertise makes a difference, as, for example, where fiscal knowledge was essential for dealing with central authorities. On the other hand, for a wide range of problems, the successes were more frequent when initiative and judgment was local; the crucial knowledge to have was knowledge about the details of the local situation. Many people have said this; Murray nails it down with statistical evidence.

Thus, a sound development policy is one that creates communications facilities not only for use by the central authorities but also for use by organizers at all levels. There is a large literature, much of it published by UNESCO, on the kinds of communications facilities that can be provided to the grass roots. There have been many experiments, in Africa in particular, with mimeographed newspapers.[22] Provincial radio and television stations have been developed in countries like Nigeria, where regional, tribal, and language differences are so great that the government has had to face up to the fact that different parts of the population want their own voice; they have had to make a strength out of that diversity rather than labeling such cultural differences as taboo. In China a system of wired loudspeakers reached almost every commune in the country. That could have been simply a way of disseminating Radio Beijing, and indeed much of the time it did replay that station. However, there were also provincial stations in various languages; and, even more interesting, the wired system allowed district officials to cut in and talk to their people.[23] The most credible voice of national policy was not a stranger talking in a strange accent from a remote capital but a local person who would be seen again the next day and who would have to answer questions about what he said.

The new technologies of communication lend themselves well to the creation of localized communications media. Even video programs can be produced locally with camcorders and cheap editing equipment. Tape recorders make it easy to get local authorities on the air repeatedly and effectively. With solid state electronics and the growing mobile radio market, production levels for transmitters are going up and costs are coming down. Photo or xerographic printing from typed or handwritten copy allows for production of local journals by technologies that are far easier than conventional printing.

The growth of local media does not necessarily split a country apart. On the contrary, the usual sequence is that the growth of local production creates a demand for networking of some kind because the local producers do not have all the material they need. The growth of local newspapers creates a need for news agencies. The growth of local radio stations creates a need for networks or syndication of tapes. No Chinese district official would want responsibility for doing a full day's radio programming for the wired network. System relations in a nation grow stronger as the needs of the local nodes become greater, not by atrophy of the local nodes.

In typical developing countries the strength of a village's structure is based on face to face contact and does not require a great deal of supplementing by new communications technologies; people see each other daily by the well or in the village square. Nowadays radio provides a steady flow of communication from the center; if the villagers are not loyal or obedient it is not because they do not know what the government is telling them. What is most lacking in existing systems is adequate communication between the village and the next level up in the political structure. Communication with the district town is haphazard and accomplished by someone taking the day to travel back and forth. Yet that is the link through which most serious business with the village is done. If there has been a crime, the village police may have to talk to the district police headquarters. The village nurse who needs medicines or advice or the presence of a doctor must get them from the district clinic. If the village teacher is to be given any help or guidance, it will be from the district education office. If there is a program for water control or pest control, coordination is at the district level. Yet little in communications facilities has been provided to link the village more closely with the nearest towns. That is usually the weakest link in the communication technology of a developing country. There is rarely a telephone or a communications transceiver in the village.

Thus the communications infrastructure required for development comprises more than satellites in the sky and television studios in the capital, or high-powered radio stations and a national elite press, but also local radio, a local press, and a variety of other devices that will allow low-level authorities to establish their legitimacy and the justice of their policies in the eyes of the people with whom they work and to communicate easily with their agents in the villages. Fortunately, technology is making local communication as well as central communication easier.

Development Is Two-Way Communication

Another critical reason why local communications facilities are important is that they make it possible for all sorts of people in the society—not just the authorities—to be heard. UNESCO publications, in which so much of the literature on communications and development are to be found, have dealt extensively with various ways of creating village-level feedback. Among these feedback devices is the listener forum, in which a group of peasants listens to (or watches) the broadcast together and then discusses it.

One of the most interesting examples of any attempt to introduce a two-way character into broadcasting is the Senegal farm radio program.[24] A program series carries a discussion among villagers of some problem they are concerned about, tape-recorded in the village. It is broadcast nationally. At the end of the program listeners are encouraged to write letters giving their own experiences or their reactions. Many of these letters are read over the air. Naturally, in the course of the forum or of the letters there have been many criticisms of the government. A minister or other appropriate official is invited to reply, and more letters are invited in reply to him. Many offended politicians, needless to say, have objected bitterly to this program. It survived only because of the personal backing of the President.

The Senegal farm radio program, like all radio talk shows, is only limited two-way communication. The tape recorder, by making it possible to broadcast real villagers from a real village, gives the rest of the audience an illusion of participation that is not unimportant despite its small size. Few people can be heard in person. But every measure toward getting the media away from the exclusive control of a bureaucratic elite helps. By all indications the program has been very successful.

Other communications devices such as the telephone or the

radio telephone are inherently two-way. Much too little attention has been paid in studies of development to the possible use of such devices at the lowest rural levels. When developing countries invest in a telephone system it is almost always with an eye to improving the service in the capital and that between the capital and other major towns. Developing countries are spending very large amounts on such programs, and for good reason.

Vast telephone projects such as those of Brazil and Indonesia are enormously useful to business development and to national defense because they permit efficient long-distance voice communication among the major centers of the country. But they are not oriented in a major way to giving a voice to villagers who want to speak to nearby villages or to their district town. To provide that kind of service economically requires a system design that has rarely even been considered. Typically, a developing country gets its engineering plans for communications systems more or less by copying what has been done in the West. That means that when a trunk is run to a population center, capacity is provided there for a switching center that may accommodate several hundred subscribers in a few years. For the moment there may be only one phone in the village, but that is regarded, incorrectly, as a temporary aberration. However, a rural village in a developing country is usually a tightly settled community, and one of the last consumer goods its inhabitants require is a telephone in each house to talk to one another. The main function of the telephone in the village is to allow the villagers to be in touch with the market town and district officials, as well as occasional contacts with nearby villages. A single community phone may be quite adequate for many years: even if it may be used only three or four times a day, it still changes the life of the village.

A system designed to be economical for such a usage pattern would be quite different from the Western model. Something more like a military field telephone or party line may serve better than a standard urban trunk-based exchange system.

There are historical analogies. In the United States in the first fifteen years of the century, farmers created cooperative phone systems among themselves, sometimes using the fence wires for transmission, and with many quality compromises to save costs. These systems spread widely and rapidly because American farmers lived on isolated farmsteads rather than in villages; they needed help in talking to their neighbors. In Germany the phone was first introduced under Bismarck as a way of getting message service to rural post offices too small to justify having a trained telegraph operator. No switchboard was provided; it was an intercom-type system. The Chinese used their wired radio system to double as a rural telepone system. Field phones could be attached to the wires for conversations among officials when broadcasts were not taking place.

A communication system, if intelligently designed, reflects the structure of the society it serves. If the society is authoritarian, it will need a downward flow. If the society is oligarchic, it will provide communication among the few. If it is a commercial society, it will need communication for business. If it is militaristic, defense considerations will dominate the design. Every institution in society needs communication; if development plans call for the growth of any institution, be it banks, factories, universities, hospitals, or cooperatives, then communications facilities of an appropriate kind are needed. A society's system of communication is a shadow reflecting the society itself.

A society that wishes to develop, in the sense of creating a population of numerous enterprising producers, needs to create communications facilities that will strengthen their local institutions and give them means to organize their efforts effectively. In an agricultural society, that means priority for rural and for two-way communications that are affordable to people other than officials and business executives. It also means priority for those simple devices that a villager can use at his initiative to meet his occasional but important needs, not just devices that talk at him.

Managing the Communications Resource

Just about all the policy issues relevant to managing the resources in a society apply to communications, too. For example, communications resources are very unevenly distributed, just as are other forms of wealth; some societies have an abundance, while others have very little. Is the gap narrowing or is it growing? The internal distribution of the communications resource within a society also raises questions—considerations of justice and equality as well as economic development.

Just like any other resource, communications can either be used for production or for consumption and enjoyment. We have here been considering their use for production and arguing that if communications are made available to those who have useful things they want to do, the resource will be used for development and growth. It is far better to think about communications as a resource than as a means of persuasion. There is, however, one important respect in which an information resource differs from other resources. It is not exhausted by use. For the developing countries, which start with enormous disadvantages in the competition for scarce resources, the inexhaustibility of information is a very important consideration.

In a world of increasingly scarce physical resources it is becoming increasingly difficult to assure every human being on earth the same amount of iron, steel, copper, beef, airplanes, automobiles, and highways that have been enjoyed by the profligate civilizations of the West. Communications facilities, however, draw only to a limited extent on those scarce resources. To enable all the world's population to enjoy music, drama or film, news, knowledge, and means of expression that modern media provide is far more feasible than to provide everyone with physical things at Western standards of desire. Indeed, in an important economic respect, information is a free good, that is, one person gains it without depriving another of it.

Developing countries stand to gain a great deal from communications technologies. By joining the most advanced international telecommunications networks they can take advantage of the latest knowledge that exists, and by extending two-way telecommunications down to the village level, they can encourage villagers to participate in the decision processes of the nation. There are, however, some problems to carrying out such an agenda.

Maintenance

Maintenance is a source of constant complaints in developing countries. Where skilled maintenance manpower is scarce, it is important to select equipment for its high reliability. That is one reason for the popularity of electronic rather than mechanical switching in developing countries.[25]

The thorniest problems of maintenance concern mechanical terminals that are widely dispersed, located in unprotected places, and used by nonexperts. The manufacture of terminals is a highly competitive field, and the devices produced vary in their reliability and maintainability. Some devices have been designed to emphasize plug-in replacement of defective modules. Probably even more important is backup by a ubiquitous service organization. In many remote locations users find it useful to have a duplicate machine on hand, since maintenance may cost more than the terminal itself.

Quality of telephone lines

The circuits that provide information and two-way communication consist of three segments: international; intracity; and city–rural. The international leg presents only one main problem, a shortage of circuits to some countries at peak hours. The quality of the signal tends to be good. A repeated comment by people from the United States or Europe who are working overseas is that they can phone more easily to their home country than to a village or town fifty or one hundred miles away.

Within the capital city, or between the port where the international circuits come in and the capital city, the quality of service ranges from poor to good but is generally fair. Though there are exceptions, in most countries one has little problem communicating within the capital city or in communicating abroad from it, at least during off-peak hours.

The critical communication problem in many countries concerns the last leg from the capital to a rural location that may be only a few miles away. In general the rural lines, and in some places the ordinary urban lines, too, are too noisy to be used for computer data transmission and may create difficulties even for ordinary conversation. In such instances, text transmission such as telex must be sent at a low data rate to minimize errors.

Capital

No matter how cheap a single use of a great communication system may be, such as a dime for one phone call or two cents for an hour of television, the cost of the total system that provides such usages adds up to hundreds of millions or billions of dollars. The organization of a modern telecommunications system requires a vast investment. The figures of investment in developed countries look astronomic to poor ones. Where can the capital come from?

No telecommunications system has ever been able to generate on its own all the needed investment capital; every one has had to rely at some time on dispersed capital raising. Even in the United States before about 1910, AT&T, though it grew as fast as it could, did not have the resources to keep up with demand. As a result, many independent phone companies emerged to serve regions not reached by the Bell System.

In some countries, where capital for the system is short, the customer has to buy his own handset. Most national monopolies, if they can, insist on supplying it, but where capital is scarce, the necessary solution may be to make the customer provide part of the equipment. In the future such a method for capital raising may be even more important. Terminals are becoming an increasingly large proportion of the total system cost. Transmission costs, as we have seen, are falling sharply. With electronic switching, switching costs will fall somewhat, too. In sophisticated systems the terminals, in the aggregate, become the major item of communications cost.

The *etatiste* ideology of many developing countries is a severe barrier to their sensibly letting customers buy their own terminal equipment, but economic necessity often overcomes ideological qualms.

Bureaucratic restrictions

The legal restrictions placed by PTTs on innovations in communications are often far more severe than the technological difficulties. In many countries it is extremely difficult to obtain a dedicated leased line; it is rarely barred by local regulations, but it may take years to have it installed. Poor countries that take restrictive postures to the flexibility in usage and equipment options are defeating their own development.

The need for international cooperation

The massive investments required in communications systems, and the economies of scale in them up to quite high levels of usage, mean that few countries can have an efficient communication system all to themselves. That applies both to small advanced countries and to large poor ones. Because no small or poor country can hope to produce all its own television programming, it is unlikely to use the full capacity of a telecommunications satellite. This situation suggests the desirability of cooperation by regional or language groupings on communications development. Unfortunately, neighboring countries are the ones that are most often characterized by rivalry or outright enmity. Even Canada and the United States, two of the more cordial nations in the world, have more tension over transborder communication flows than over almost any other issue. Kenya, Uganda, and Tanzania formed a telecommunications union with a shared ground station and a single development effort, but first, under Idi Amin, Uganda's membership became a formality, and then Tanzania and Kenya found themselves with a closed border and fighting between them. In 1977 the union was dissolved.

One may hope, nonetheless, that small or poor countries will often cooperate in creating regional and global consortia for domestic as well as international communication. No small or poor country has enough communications traffic to justify its leasing large numbers of point-to-point circuits on a satellite, not to mention having a satellite of its own.

Most of the INTELSAT circuits go to and from Europe, America, and several Pacific Rim countries. Fewer links connect developing countries to each other. The fact that two developing countries do not have a circuit between them does not mean that they cannot communicate. They can do so via a third country, but that requires a double hop, and extra costs.

There are means by which all nonvoice traffic to or from a country can be loaded onto one or a few circuits. In a multiple-access satellite system, all traffic is broadcast on a single frequency. Computers at the various ground stations pick out only those messages addressed to the host country. All international (nonvoice) message traffic to almost all countries could be handled on a single phone circuit apiece.

In the long run, the costs of electronic communication of text can be brought down to very low levels, well below that of the mails. However, to achieve those economies will require use of batch transmission and message switching and will also require sharing by *many users* of high bandwidth circuits and of satellites.

Only the largest countries have that many users among their own citizens. Poor or small ones may have to establish cooperative sharing arrangements.

Islands of modernity

Developing countries are dual societies. Side by side with abject poverty and primitive peasant life one will find research laboratories, airlines, sophisticated weapons, wealthy cosmopolitan elites, world-class hotels, symphony orchestras, and modern hospitals. One can agree that some of those expressions of modernity are horrid and abusive and should be abolished; yet others are the very hope of progress. During rapid development, with occasional dramatic leaps over stages, it is inevitable that some parts of the society will be left as they have been for centuries past while other parts move into the twenty-first century. One certainly would not wish to deny to a developing country the possession of research laboratories, factories, and hospitals up to the standard of the best in the world. Such islands of modernity will need communications systems as modern as those anywhere.

Though developing countries can and should use some super-sophisticated technologies, they often can do so only in a few locations. The difference between an underdeveloped and an advanced country may not be so great in the kinds of things done in the top intellectual and organizational centers but rather in the extent of diffusion of advanced methods. The econometric analysis or computer facilities in the top university or central planning office of a less developed country may not differ much from that in a developed country, but the back up and numbers of such centers may be far less.

The implication of this fact is that in poor countries the benefits of these few sophisticated activities must be diffused by telecommunications. An advanced country can afford to have many good libraries, computer centers, medical centers, or teacher-training institutes. A developing country may have only one of each and therefore a much greater need to put it on line to the whole country. Training teachers by television, advising medical technicians by telephone, and backing up local administrators by on-line services may be even more important in a developing country than a developed one.

Fostering Development in an Electronic Age

One of the vexations of underdevelopment is that it is all of a piece. It would be distressing enough if the only problem in underdeveloped countries were that people were poor. But along with that, life is short, medical care is inadequate, education is lacking, security is poor, governments are corrupt and inefficient, progressive political movements are disorganized, and technological competence and entrepreneurial motivation are scarce. All the problems come together. Progress cannot be made on one front at a time; it requires what Mao Zedong called a "great leap" and what Walt Rostow called a "take-off" on all fronts at once.

It has not always been recognized that the evils of underdevelopment form such an interlocked package. In the eighteenth century there was in Europe a romantic notion of the noble savage. He may have been poor and lacking in science and culture, but he was somehow purer, more upstanding, and more understanding than his civilized successors. That romantic illusion is fairly well dead today except in the form of occasional self-justification by the less successful governments in poor lands who boast of some intangible merit as a way of diverting attention from stagnation. In fact, all the evidence supports the more common-sense notion that where people are richer they are happier.[26] Where they have more access to the world's culture and knowledge, they make more and faster progress in both wealth and welfare.

Honesty in facing those facts is important, yet one is reluctant to assert them. Pride is a condition of progress. Bald honest assertion of how bad, corrupt, and incompetent things may be in a poor country slides over easily into self-destructive negativism. Myths and illusions do play a role in human progress. Max Weber has analyzed how, at an earlier stage in European development, the confidence of the Calvinists in divine personal election contributed to their effectiveness as entrepreneurs.[27] So too in the less developed countries today, ideologies about cultural uniqueness and special national merits play a curious double role. Often they are reactionary myths that serve to obstruct progress, but they are also often part of an ethos of pride and self-confidence that serves in fact to promote effective action and therefore to advance change and progress. Under some circumstances national traditionalism may, in Marx's phrase, contain the seeds of its own destruction by fostering a vigorous drive that leads to development and progress.[28]

The situation, in short, is not one to make simplifiers happy. We can start right out by dismissing all the usual clichés about the protection of national cultures, and equally dismiss their contraries. To achieve what makes people happy and what they choose for themselves when they have a chance—not necessarily what their governments or ideologists tell them they ought to want—underdeveloped countries need a complex mixture of things. They need a positive feeling about the greatness of their culture and their past. They need to have new ideas translated into styles they are comfortable with; even in successfully modernized countries like Japan, the cultural envelope may remain very distinctive indeed. On the other hand, they need and want enormously rapid change; the change must be much more rapid than anything experienced by the countries that modernized in the eighteenth and nineteenth centuries. If it is not, underdeveloped countries will never catch up. They need to imitate on a grand scale, for that is the way most cultural change takes place. Any country doomed to invent everything for itself would be stagnant indeed.

In summary, opportunities for using communications to accelerate these changes are there, waiting to be picked up by any country that wants to use them. Three important improvements in technologies of communication are making it easier to do so. Low-cost cameras and editing equipment, satellites and low-cost ground stations, along with parallel developments in publishing, are making it cheaper and more convenient to get mass media to the population; the cost of entry into television broadcasting in particular is falling. Second, international data networks are eliminating geography as a barrier to access to the world's best information resources; users can be on line to them from anywhere. Third, satellites and solid state electronics makes local and two-way communication more feasible, so that even the remotest villages can be brought into the national system.

The barriers to taking advantage of these opportunities are political and ideological. Countries are often reluctant to open up their borders to communication. They are unwilling to engage in joint communications development with others. They are afraid of providing diverse and sometimes discontented groups with easy two-way communication. They are offended by the notion that they should give free rein to creative people who rarely have much sympathy for the authorities.

For a country to develop its own productive capability in communications, the most important single requirement is that there be a lively and encouraging atmosphere for the practitioners of communications enterprises and the arts. To develop movie making there must be movie theatres, a movie-going public, and a political atmosphere that encourages imaginative, creative motion-picture people. Clamping down on the flow of movies and censoring movies that people want to see does the opposite. To develop its own journalism, a country needs lively, controversial newspapers with avid and politically interested readers. Barring interesting material from outside sources has the reverse result, as does the mistreatment of journalists.

Self-defeating political efforts to develop communications in such a way that only the "right" messages are communicated and not the wrong ones, and in such a way that it is all done indigenously, and that the use of the facilities is kept out of the hands of anyone who would misuse them, are what stop the development of communications in most countries, not any external exploitation, nor even lack of capital, though the latter is certainly a constraint.

Neither in communications nor in any other field can a country cursed by poverty, lacking trained and skilled personnel and with age-old obscurantist traditions, change everything at once, to become rich and modern overnight. It can take advantage of coming late to learn what others have done before, skip stages, and move fast. But if it wishes not to squander its limited resources, it must apply them for those special uses wherein they have a comparative advantage and take full advantage of what can be obtained more cheaply on the world market; that applies to communications as well as to other resources. Also, no country, and particularly not a poor one, can develop its resources rapidly with only a single or a few centers of initiative. If it wishes to develop its communications capabilities, it must encourage creative people throughout the society to communicate freely.

Developing countries should insist on the freest possible access by telecommunications to the best available information resources from anywhere in the world. It is in their interest to oppose all restrictions on the free flow of information. Copyright, security restrictions, and commercial restrictions on the transfer of technical information all check the progress of developing countries. Rapid transfer of technology is facilitated by the maximum possible flow of travel, messages, literature, and on-line interaction across the world's borders.

Dependence occurs whenever advanced countries possess know-how and techniques that developing countries are not able to acquire for themselves at will. Independence is therefore promoted by unrestricted free flow of information between countries so that the developing country can acquire for itself whatever intellectual and cultural products it desires at the lowest possible price. The freer the flow of information, the wider the developing country's range of choice and the sooner it can acquire for itself the ability to produce the same sort of information or programming at home.

The conclusions presented here are rarely accepted by government, including the governments of underdeveloped countries. Governments everywhere have an inherent bias in favor of controls, restrictions, and provincialism. The business of government, after all, is to govern; a rare government indeed is the one that recognizes when it is hurting its own people by its regulatory actions. It is a sad but undeniable fact that the largest part of all the oppression, brutality, and sadism in the world is done by governments against their own people. And the governments of the less developed countries often suffer from underdevelopment, too. So the fact that few developing states have encouraged or even allowed a free flow of information is no test of its value for them. Moreover, the judgment of international organizations on the value to development of the free flow of information is likely to be no better than that of the governments that compose them and instruct their representatives to them. International organizations are likely to misinterpret the needs of the less developed countries, perceiving them through the distorting vision of government officials.

Chapter 10

Advanced Communications and World Leadership

Communications aid the development of nations in three primary ways that we will explore in this chapter. First they serve to project national power. They are also important in the conduct of foreign trade. And they contribute to the efficient distribution of activities. Communications are particularly important for the transfer of such capabilities to the less developed countries, but they serve such purposes in highly advanced economies as well.

Telecommunications and the Projection of National Power

A nation's influence in the world is stronger if it advances well in the development of its arms, industry, food production, public health, and education. The same is true for developing its communications. Most particularly in this era of low-cost global communication, if communication services can be accessed at approximately equal cost from anywhere, then people everywhere will wish to use those facilities that are most efficient, cheapest, and best developed. For example, if one uses a library by sitting at a computer terminal rather than by traveling to the books, and if distance does not enter into the information retrieval cost, then one will use the best and cheapest library, wherever it is. There was a company that sent messages between two plants in the same town in Germany via its switching computer in the United States. That switching computer could be located anywhere; the company chose to put it where the best hardware, maintenance organizations, and software programmers were to be found, and (even more important) where the government regulations did not restrict it from doing what it wanted to do. For that company at that time, such a location happened to be the United States. In Europe many U.S. and other globally far-flung companies have put their communications hubs in London because it has the least restrictive communications policies. This has contributed to making London a major center for commercial transactions.

The geopolitical advantages of communications leadership are not new; they were understood by both Britain and Germany in the years before World War I. The British recognized early that the nation controlling the undersea telegraph cables would, because of its influence over the flow of news, have an advantage in trade, military strategy, and politics. British capital took the lead in laying cable across the Atlantic, and by 1914 the world's cable network was mainly a star with England as its hub. The flow of news between the United States and the European continent went through a British bottleneck.[1] This control of world communication proved a major asset to Britain in World War I.

After the war, the formation of RCA with the cooperation of General Electric, AT&T, and Westinghouse at the urging of the U.S. government (despite antitrust laws) was motivated by the government's fear that the British would gain a similar control over the new long-distance technology of radio. RCA was formed in 1919 to prevent transatlantic and maritime traffic from being monopolized by the British Marconi company.[2]

The technology spectaculars of the postwar years that have had the greatest political impact have in many instances been on the edge of communications. With the Sputnik satellite in 1957 the Soviets changed the image of a whole generation as to their competence and technical prowess. A typical poll in 1958, the year following the successful Russian launch of Sputnik and the United States' failure to launch a much smaller satellite of its own, would have shown that American confidence in its scientific, technical, military, and educational institutions was shattered. It took America's sending a man to the moon to overcome that image.

It would be hard to overestimate the contribution of the space program in the 1960s to America's image in the world as a leader. The man on the moon was a psychological triumph. The benefits to the United States of the space program, moreover, went far beyond the immediate goals established by President Kennedy. The intelligence from satellite reconnaissance on Soviet missile pads was invaluable. For a long time, other countries had to turn to the United States for meteorological and earth observation data and facilities.

The prestige and other advantages that the space program gave to the United States are, however, waning assets. The American monopoly on various space technologies did not last long. In 1974 the Soviets lifted a satellite into geosynchronous orbit. The European Space Agency (ESA) successfully launched a geosynchronous communication satellite in 1988; the Japanese plan one for the 1990s.

Export of Telecommunications Hardware

Besides prestige and strategic advantages, another motive behind the development of international communications is to lay the foundation for export business in communications equipment. If one reads what is written in various countries about international communications policy, one might conclude that often the tail of equipment exports is wagging the dog of communications services. One finds the authorities in each nation simultaneously setting idiosyncratic standards for communications equipment so as to protect the home market for their domestic manufacturers (as in the case of the French color TV system) and at the same time trying to open up trade in communications equipment so that they may sell abroad. In 1977 officials of Britain's engineering labor union blamed the post office for using a unique British telecommunications system that no other country would buy; that, they said, resulted in a loss of 10,000 jobs and a prospective loss over two to three years of 20,000 more in telecommunications manufacturing, an industry in which Britain, according to the union, was once the leading exporter.[3] As is usual in self-serving debates about trade, the interested parties want it both ways: uniqueness in order to protect the domestic market and compatible products that can also be exported.

In every country the communications industries are under public pressure to expand their sales abroad. From an Olympian perspective these struggles by competitors to protect their own turf and expand into others' are a spectator sport with more noise than movement. What is significant about the struggle is the testimony it gives as to the national advantage that can be gained by the intensive development of communications.

Facilitation of World Trade

Whatever may be the interest of particular nations in communications as a means for expanding their equipment exports, more important for overall welfare is the contribution that communication facilities may make to the expansion of overall world commerce.

Obstructions to communication are one of the chief impediments to world trade. Off at a distance, a buyer or seller has trouble keeping informed of the details of a market abroad and its short-run fluctuations. Even after a purchase order has been negotiated, documents must be transmitted confirming the funding and the shipping arrangements. Information may be wanted about the location of the shipment at any given time and about delays en route. Currently, communications breakdowns and delays in receiving documentation are a major burden to trade, particularly since the information alone is often not enough to allow action to go forward.

In international trade, since the seller is in one jurisdiction, the buyer far away in another, and the goods perhaps in a third, it has long been customary that the goods themselves are not released by the shipper until he receives a legal evidence of payment, nor will the banker release money in payment without a legally certified bill of lading. An uncertified check or letter or telex message will not do. The legal formalisms of stamped, signed, sealed documents still hold.

To help meet the need for documentary evidence while speeding up the process, facsimiles of legal documents are beginning to be accepted as evidence. International voice telephone is also important to traders to help them achieve the interpersonal persuasiveness and confidence on which business transactions depend. Telex and fax messages are important to business especially in those areas where voice phone service is unreliable, where different time zones make conversations difficult, and where documentation is important. Trade is a strong predictor of international telephone, telex, and telegraph traffic.

Computer-based message handling systems that provide store-and-forward capability, message switching, and packet switching will not only bring the cost of message delivery down but also make possible flexible kinds of business interaction that are now difficult at a distance, including written teleconferencing and electronic mail. Low-cost data communication supplemented by voice may begin to give the trader a level of supervision over his transactions and over his goods abroad approximating that which he now has in domestic transactions. Presumably, in time, electronic documentation will be accepted in place of the hard-copy documents now required, and with that days and weeks of delay will be eliminated. The expansion of communication and the expansion of trade go hand in hand.

Trade in Knowledge

Besides its contribution to facilitating trade in goods, improved communications also give rise to an invisible trade in information. The importance of invisible services in world trade has been growing. Such invisibles amount to one quarter of world trade. The United Kingdom has relied on invisibles to offset a trade deficit for 200 years.

Information, of course, is only a small item among these invisible services. Remittance by immigrant workers to their homelands, government grants to foreign countries, and insurance and shipping charges are much larger. But charges for information are not trivial. There are also commissions from business services, payments for publications and TV and movie rentals, travel bureau commissions, and telecommunications charges.

Once the costs for access to information services become largely distance-insensitive, new opportunities for international trade in information open up. Any institution that produces information and stores it in a computer can publish that information by allowing anyone else with a communications terminal anywhere in the world access to it. As we noted previously, the information creator will often prefer to have a middleman handle the archiving, indexing, billing, and so on, and so the vendor may often not be the original institution. But either way, remote access to the information is possible.

International Pressures on Communications Policies

Improved communications may remove burdensome constraints from the topography of human activity, but makers of public policy are rarely happy to see that happen. The political process generally protects established interests against social change. For example, the idea of building new towns was very popular in the 1950s and 1960s when the combination of a baby boom and a housing shortage lent support to planners who argued for the greater humanity of life in green belts with lowered densities; they could point out that improvements in communications made it less necessary for people to congregate all together. Tama New Town was established outside Tokyo, Reston outside Washington, Milton Keynes outside London. But as soon as people and businesses really began moving to exurbia, leaving unrented offices downtown and cutting revenues of the cities, government policy reversed. In England, for example, there are now to be no more new towns started and government efforts go into "rescuing" the dying metropolis.

So too with the effects of improved international communication. Authorities who in principle favor free flow of information and praise cultural exchange quickly reverse course when these flows have an impact on established domestic institutions. That happens more and more often. Large-scale, low-cost global communication increasingly affects domestic interests and constrains the alternatives in domestic policy. No country can impose effective privacy legislation without considering how it can be evaded by data sanctuaries. No country can open its political system to contested elections without recognizing that parties will be formed committed to and backed by the major global ideologies. No country can expand television broadcasting or have a press without at least considering the economics of accepting advertising, which will come in large part from multinational firms with something to sell. No country can impose censorship on pornography without expecting extensive evasion by foreign literature. No country can impose political censorship without knowing that short-wave broadcasting will penetrate it. No country can give domestic companies a monopoly on information services or computing without recognizing that users will have work done abroad to evade the monopoly. No country can enforce royalties and copyright on popular music if foreign broadcasters play it and young people have tape recorders. No country can decide how to allocate spectrum or where to put what kinds of satellites without compromising with the demands of the ITU.

Cosmopolitans may not worry about such restrictions on the effective power of sovereign states, but political decision-makers do. A decision as to how to structure a nation's communication system can no longer be a domestic decision alone. Communications are decreasingly confined within national boundaries except where governments enforce that limitation brutally. Most communications problems that regulators try to solve on the domestic scene will have to be dealt with internationally, too, with all the added difficulty there is in resolving conflicts among nations.

One area in which international cooperation is especially beneficial is in the operation of satellites. The capacity of standard satellites far exceeds the needs of most countries, and they are expensive. Moreover, the optimal satellite design for broadcasting, for data, and for telephone differs in regard to power and also frequencies. Therefore, it makes much more sense to have several specialized satellites, like a Marisat, an Aerosat, a high-powered broadcasting satellite, a low-powered telecommunications one, each serving many countries, than to have one multipurpose satellite serving one country or a few. But this increases the complexity of international cooperation, for more countries must work together on each satellite.

In almost every country (the United States being only a partial exception), one of the key issues on which the conflict between new international communications technology and domestic policy is being fought is that of the PTT monopoly. Post offices have always been monopolies. The justification they offer has been that they are obliged to provide a universal service to the whole population and that means service to some elements of the population at a loss. If they are not given a monopoly, competitors will "skim the cream," providing cheaper service to the highly profitable parts of the market and leaving the losing part to the PTT.

In the earliest days of the postal service the subsidized part of the service was the government's message traffic. Companies were given a franchise to carry official government mail, and at the same time given a monopoly on carrying private mail for citizens. For the private traffic they could charge what they wished, and thus they were assured financial support for carrying the King's mail.[4] After the Rowland Hill postal reform of 1837 in England, which established uniform national rates (a reform that was soon copied elsewhere), the cross-subsidization was borne by the metropolitan customers on whom the system made money, for the benefit of remote rural populations who were served at a loss. Monopoly and that kind of cross-subsidy was justified as creating national unity and national development. Today, the same argument is usually made by telephone systems about cross-subsidization between the modest home telephone subscriber and the business service subscriber. The claim is made in most countries (and often validly) that the high rates charged business keep the rates to ordinary subscribers down. In many countries heavy rates for international customers also benefit local subscribers.

Increasingly, this monopoly, which has become an article of ideological faith among PTTs, is threatened by new technologies of communication, and particularly by those technologies that do not respect boundaries. The monopoly depended upon the fact that any system of universal communication required a large transmission plant within the country. Post offices, vehicles, and a large corps of delivery personnel were required for a national mail service. Broadcasting stations and spectrum were the privileged plant of a few licensed broadcasters or of the state broadcasting monopoly. A vast wire and switching plant was necessary for a phone system.

In some aspects of modern communications systems, however, that pattern is changing. Short-wave broadcasting was the first example. With a relatively modest technical plant, it is possible to reach people in other jurisdictions; an individual's ticket for participation in the system is buying a receiver. To control reception, governments have to control not just the central transmission organization but also the widely dispersed private receivers.

What has been possible for some decades in the case of voice broadcasts by short-wave radio is becoming possible with the aid of satellites and computer terminals for many kinds of communications. The limits are the economic ones of the cost of the terminal compared with the cost of traditional means of delivery. We have already seen that the direct satellite transmission of television pictures to homes, which requires much bandwidth and expensive terminal equipment, is competitive with conventional television only under very limited circumstances—and yet even that has frightened governments concerned with their sovereign authority. Other sorts of transmission are clearly going to be so economical as to justify the alarm of PTTs about the challenge that extraterritorial activities pose for their monopoly.

Sovereign nations can, of course, do what they wish, and some will deny their citizens access to the emerging global facilities, just as some prohibit access to certain foreign printed material. Such laws are only partially enforceable, and they have their price in efficiency and morale in the country that tries to enforce them. The destruction of the monopolies are, however, no more inevitable than the destruction of censorship; they are both archaisms that some countries will try to preserve.

The United States is a special case. Except for the carriage of first-class mail by the postal service itself, there is no legal communication monopoly in the United States. Electrical communications are handled by private companies. But even in the United States there is unresolved ambivalence. Regulatory agencies recognize a responsibility for protecting the existing institutions that are providing service to the public; they are unwilling to see carriers start to lose money and let their services decay. Regulators are also nationalists and are unwilling to see foreign communications services enter into significant competition in the domestic field.[5]

The ambivalence in American policy is illustrated by the difference in the FCC's international and domestic satellite policy. The policy of international monopoly which the United States adopted regarding INTELSAT was hardly congruent with the policy of open-skies competition that was adopted for domestic service, particularly when the two kinds of service are technically indistinguishable. The rates charged for these indistinguishable services ended up being very different indeed. We can find differences of up to 10 to 1 between rates quoted by the international record carriers and those quoted by competitive American domestic satellite entrepreneurs for comparable services.

As for its acceptance of certain international restrictions on communications development, the U.S. position is not simply one of libertarian principle compromised by pusillanimity and a desire not to offend. That is part of the picture, but would that it were that simple. American behavior is also that of a nation-state seeking its own perceived self-interest. That self-interest has sometimes been served by barring foreign competitions in situations similar to those where Americans would also like to compete abroad; American authorities under political pressure have tried to stem the flood of Japanese, then Taiwanese, then Mexican radios, TV sets, and VCRs.

On the other hand, the U.S. interest is more often served not by traditional protectionism but by encouraging the most rapid possible innovation and improvement of the world's communications systems. The United States is the leader in the development of satellite technology; it is the major exporter of satellites and the equipment needed to use them. The United States is also a world leader in manufacturing electronic equipment, particularly computers. As such, it stands to lose by widespread policies of protectionism. America is also the world's largest source of commercial and private travelers; so it stands to benefit if communication services operate well in other countries.

The United States faces a classic policy problem in international relations. Can it persuade others that communication is not a zero-sum game but that in the end all will benefit from the free flow of global intercourse? As a principal beneficiary, the United States finds it hard to persuade others to forgo the apparent, even if illusory, short-run advantage of restricting foreign access to their communications flow.

The American Role

The policies that the United States adopts regarding the organization both of its own communication system and of international communication are of course important to Americans, but they are important also to the whole world, for there is every reason to believe that for the foreseeable future the United States will continue to be at the leading edge in this field.

At least it has all the opportunities for continued leadership. It has a telephone system that works well—unlike that in most countries. American companies not only manufacture most of the world's advanced computers but also have the largest market for them. With the present rate of diffusion of terminals, computers could be approaching the stage of more or less universal penetration in the 1990s—the stage similar to that which the telephone reached in the United States in the 1950s. Finally, it is in the United States that the technique of packet-switching data has been developed. A universal switched system, computers, data networks, and satellites are all key elements of the present communications revolution.

America has a profound interest in maintaining this lead. Such a lead may do for another generation what American technology and mass communication have done in the recent past. American national interest has clearly been served by the spread of American films, books, art, magazines, business firms, and TV serials. That can be said without for a moment accepting the simplistic zero-sum thinking which sees world leadership in communications as harming others and serving imperialism. To recognize the gain of one is not to assume a loss by another. A mutuality of interest exists between an innovating nation that advances communications technology for its own reasons and a receiving nation that buys advanced equipment from the former and thereby bootstraps itself into technical sophistication.

And yet American communications policy has not been designed to maximize the American lead. As we have seen, an enormous burden of restrictions seems likely to delay or obstruct the development of international communication services potentially valuable to the United States. Technical developments that could add to American trade and prestige and further the development of cultural relations are in many instances illegal under present regulations, or made so expensive as to be impractical.

While the most severe restrictions are imposed by foreign governments, it has been hard for the United States to effectively combat such restrictions, partly because American policies have in many instances also been restrictive on international traffic.

Although commercial communications services have been and probably will continue to be the main channel for American leadership in the development of global communications, government activities have also played a very important role. There are, of course, those government activities in the electronics field that are known only by leaks. The electronics that permits a cruise missile to travel unmanned for thousands of miles and then recognize its target and hit within 100 feet are obviously extraordinary, even if those of us on the outside do not know what they are, and even if their cost is such that they are not going to enter any consumer product in the near future. So too are the electronics of vehicles that are sent to explore the solar system.

Even secret military activities have a considerable diffusion effect, for secrecy basically does not work.[6] Quite apart from espionage and leaks, when something has been demonstrated in one place in modern science, it is only a matter of a fairly short time before scientists elsewhere can figure out how it was done. And the fact that it was done by one superpower is a driving incentive for other powers to do the same. Sputnik created the U.S. space program.

Thus, despite the highly protectionist and restrictive attitudes that prevail in both the United States and other countries, the potential of useful applications of telecommunications technology is likely to be realized. The restrictions will be only delaying factors.[7] The world is a competitive place, and competition has its inexorable logic. As argued before, those countries that try to protect themselves from international competition in data services will find that other countries are gaining advantages in productivity and economy by means of superior computer telecommunication. For their own advantage they will need to lower the bars. What country will want to deny itself access to valuable data bases or to meteorological or resource information? Once the Pandora's box of technological competition is opened, each country would be well advised to move fast to participate as an efficient user and as a producer of international data communication.

Many countries, perhaps even the majority, will initially be restrictive of innovation. Regardless of how shortsighted the regulations by some nations may be, regardless of how troglodytic the discussions in some international organizations, still those nations that seize the new technological opportunities will advance faster than those who do not. The processes of competition between nations will in the long run make mercantilist policies nonviable. If in the 1990s the choice for small underdeveloped countries is between trying to build all their own information services, on the one hand, or to link up on-line to the Library of Congress, the National Bureau of Economic Research, the EOSAT earth observation satellites, and perhaps by that time to the Soviet Gosplan planning models, clearly those who choose to use all the available resources are the ones who will progress. Thus, in the end the odds are in favor of the successful emergence of global communications services.

Improved telecommunications can contribute so much to productivity, to scientific and technical progress, to public health, to international security, and to human enjoyment that it will only be a matter of time before reluctant governments see the desirability of moving in the way that the most progressive governments have already moved.

PART III

ECOLOGY, CULTURE, AND
COMMUNICATIONS
TECHNOLOGY

Chapter 11

The Ecological Impact of Telecommunications

In a world of scarce resources, thought is pleasingly abundant; like air, it is a free good. But the means for communicating it are not equally abundant. Travel is a wanton drain of energy. The post and telephone are much cheaper, though not always cheap enough for the poor. Mass media are inexpensive when their audience is large—the secret of their success. Individualized communication and group communication are falling in cost toward near the low level of the mass media. Communication, in short, is one of the good things in life that can be had without straining the world's scarce resources. In communication we are very far from the limits of growth.

That fact underlies most of the key developments whose social implications we examine in this book. The falling cost of electronic logic supports the trend toward individualization. The growing abundance of bandwidth in transmission and better management of the electromagnetic spectrum creates the technical opportunities for small-group communication. Satellites and fiber links are making costs more distance-insensitive and underlie the internationalization of communication. Telecommunications are also likely to promote urban decentralization—the movement of business and cultural life from overcrowded cities to exurbia.

The relationship of computers and telecommunications to regional planning and decentralization has been extensively discussed in Europe. In France, Japan, Sweden, Great Britain, and many other countries there is a widespread desire to check the growth of the capital city and to disperse more of business, industry, and cultural activities to smaller urban centers and new towns. A number of authorities, for example Peter Goldmark, argue that with adequate communication facilities, rural life can be made attractive enough to check migration to urban metropolises, and indeed attractive enough to draw people out of the metropolises. Such decentralization, it is argued, could produce vast savings, for the capital required to enable millions of people to cohabit the same few square miles of ground is enormous. One must dig deep under the ground and build steel and cement towers above it. Transporting food and other consumables into cities and removing waste are complex and expensive operations. On the other hand, there are costs of decentralization, in roads that must be built, for example. But certainly if improved communication permitted life to be attractive either close-in or far-out from cities, as economic considerations dictated, and permitted business to be conducted efficiently from either place, then activities could be located wherever economically optimal in terms of transportation and other resource costs. Removal of one constraint on location cannot help but bring costs down as locations are picked to satisfy the remaining constraints.

Communication and the Pattern of Urban Settlement

The available means of communication and transportation affect the pattern of human settlement. If people can deal with one another only by walking, they will settle much more closely than if they can transmit messages or phone one another. Earlier we saw what a profound impact a change in communications technology can have on spatial patterns. Before the telephone, doctors, for example, had to live near their offices to be readily available when needed; the office was, in fact, typically in the doctor's home. The telephone allowed many doctors to separate home and office and put the office where it was convenient for patients to visit.[1] Once the telephone was available, business firms could move to cheaper quarters and still keep in touch. A firm could move outward, as many businesses did, or move up to the tenth or twentieth story of one of the new tall buildings. Instead of a checkerboard of different specialized neighborhoods, the new urban pattern became one of a downtown containing a miscellany of those commercial and marketing activities that needed to be accessible to clients and customers, as well as a growing set of satellite downtowns, for more convenient shopping and services, and the exiling of those activities that needed little outside contact (like manufacturing) to peripheral locations.

The telephone favored the growth of skyscrapers in several ways. As mentioned above, without telephones human messengers would have required too many elevators at the core of the building to make it economical. Furthermore, telephones were useful in construction; the superintendent at the bottom had to keep in touch with the workers on the scaffolding, and phones were used for that. As the building went up, a line was dropped from the upper girders to the ground.

Thus, contrary to the simple notion that the telephone led to the growth of suburbs and to urban sprawl, we find that it initially led to concentration downtown. Moreover, the movement out to residential suburbs actually began in America in the decade before the telephone and long before the automobile; it was based on the street car.[2] The conventional assumption is that the automobile and the telephone—between them—were responsible for the vast growth of American suburbia and the phenomenon of urban sprawl is challenged by the reverse proposition that the telephone also made possible the skyscraper and an increased downtown concentration.

In these contradictory ways, the telephone helped create regional agglomerations. As Jean Gottmann stresses, the telephone favored megalopolis, not antipolis.[3] A megalopolis (Gottmann's term) such as the Boston-to-Washington corridor is not an undifferentiated sprawl of medium-density settlement. It is a highly differentiated structure with numerous centers and subcenters having complex interrelations. Other commentators have seen this development as the destroyer of great urban cultural centers; Gottmann disagrees.

One can find quite early predictions of the process that we now call the formation of megalopolis. In 1902 H. G. Wells forecast centrifugal forces on cities that may lead "to the complete reduction of all our present congestions."[4] A pedestrian city, he said, "is inexorably limited by a radius of about four miles, and a horse-using city may grow out to seven or eight." With street railways the modern city thrust "out arms along every available railway line."

> It follows that the available area of a city which can offer a cheap suburban journey of thirty miles an hour is a circle with a radius of thirty miles . . . But thirty miles is only a very moderate estimate of speed . . . I think, that the available area . . . will have a radius of over one hundred

miles . . . Indeed, it is not too much to say . . . that the vast stretch of country from Washington to Albany will be all of it "available" to the active citizen of New York and Philadelphia.[5]

Wells saw the telephone as one factor fostering this development; there was no reason "why a telephone call from any point in such a small country as England to any other should cost more than a post-card."[6] Yet Wells, like Gottmann later, emphasized that urban sprawl did not mean uniform density. Shopping and entertainment centers would continue to make downtowns, while people in some occupations would prefer to move out to the country and work by phone from home.[7]

A *Scientific American* article in 1914 played on similar themes, but with special stress on the picture-phone as likely to make dispersion possible. "It is evident," it began, "that something will soon have to be done to check the congestion" of the city. "The fundamental difficulty . . . seems to be that it is necessary for individuals to come into close proximity to each other if they are to transact business." The telephone and picture-phone, the article argued, would change all that.[8]

The telephone not only contributed to the break-up of single-trade neighborhoods but also tended, at an early stage of its development, to encourage the stabilization of "good" neighborhoods and business districts and their separation from areas of decay. Phone entrepreneurs, when they were getting started, like cable television entrepreneurs today, could not afford to wire the whole city at once. They preferred to lay their lines where they could expect to recruit many subscribers, that is, in affluent neighborhoods and business districts. Having a concentration of potential subscribers in a confined area was economically advantageous to the utility. Shifting and deteriorating neighborhoods were not good for business as capital-intensive infrastructure. Zoning of a city helped in planning for future services, so the phone companies (along with other utilities) became supporters of the zoning movement. The Department of Commerce's zoning primer of 1923 states: "Expensive public services are maintained at great waste in order to get through the blighted districts to the more distant and fashionable locations."

Zoning, along with other efforts at urban planning, became popular around the turn of the century; it actually began in New

York in 1916. In the crucial second decade of the century, phone companies were one of the main sources of information for new urban plans. The Bell System collected large amounts of neighborhood data on the population trends in the city, its businesses and neighborhoods.[9]

In the nineteenth century, well before the invention of the telephone, directories of cities were already being published. From the beginning of switched telephone systems it was recognized as important to give subscribers a list of other subscribers. This list was particularly important if subscribers were to be called by number. These listings were originally printed on a card to hang by the phone. By 1897 the national phone directory had become too big a book and gave way to local directories. As telephones became universal among businesses and later the public, the phone book became the most widely used city directory. It had to be issued often to keep up with growth and change; it was available everywhere; and it was free. The inclusion of yellow pages made it useful for many business purposes, and it became the basis for much canvassing and sampling. Other city directories, sold for profit, were forced to provide more information to survive. For the ordinary citizen the telephone book became the standard way not only to find phone numbers, but also mailing addresses and business locations.[10]

In a variety of ways, then, the telephone has facilitated the coordination of a complex urban system. When the telephone was fifty years old, Arthur Pound in an anniversary volume noted how the life of a city could be brought to a crawl if the phones all suddenly stopped working.[11] The trains, the produce dealers, the hospitals would all grind down. It is hard to conceive of a metropolis running its myriad functions well without extensive use of the telephone. Some cities function with much less use of the phone than we take for granted. Beijing, Calcutta, and Moscow are large cities with very limited telephone service by American standards. Yet they function. But it hardly need be doubted that the way they operate suffers severe restrictions from their limited use of the telephone and would suffer worse were there none at all.

The telephone also shaped the countryside, together with the mail and the automobile. Together, they led farmers to do their banking, buying, and selling in larger, more remote centers rather than at the nearest rural village, and led villagers to take their trade to better stocked, more developed but more distant centers. Rural free delivery, in particular, favored the growth of mail-order business and reduced the need for farmers to come to the local post office. Parcel post was established by Congress in 1912 over the vigorous opposition of the small towns—the backbone of American democracy—which feared their commercial role would suffer by the increased competition of mail-order vendors.[12]

Most rural phone systems were at first tiny local networks providing village communication only. By the second decade of this century these systems were being rapidly interconnected with the national network. They no longer just linked farmers to each other or to the nearest small town but connected both to the national urban system. It was no longer necessary to take a trip to deal with an urban center. One could substitute a quick phone call.

The Communications/Transportation Tradeoff

The notion that improving communications facilities may reduce the need to travel is not a new one. Within three years after the invention of the telephone in 1876, the London *Spectator* predicted that the new device would substitute for personal meetings.[13]

However unrealistic the early prediction of reduced urban congestion may seem to us today, there was good empirical evidence behind these speculations. It was noted that the pattern of activity of traveling salesmen was changing. They continued to call on customers, but less often in person; they could phone between visits, and for some purposes could prepare for a visit by a phone call.[14] Some railroads claimed to be able to feel a resulting decline in their traffic.

With hindsight we know that travel, transportation, and traffic congestion has grown in this period of improved communication as never before. Indeed, even those who in the early years of the century were anticipating the replacement of travel by communication could not help but notice how delayed the process was. Herbert N. Casson, an early chronicler of the wonders of the telephone, claimed in 1910 that "slowly and with much effort the public was taught to substitute the telephone for travel."[15] Yet in this early literature we do not find the counterproposition that is often put forward today, namely, that the telephone increased relationships with people at a distance, thus leading to an increase in travel.[16]

A careful review of the relation of the telephone to transportation would show a number of complex directions of causality. The telephone, as well as the telegraph, facilitated better coordination of trains, traffic, and deliveries, making service prompter and more reliable. One notable result was an increase in the remote marketing of perishable goods.[17]

We cannot draw any single-valued conclusions about the relationship of the telephone to transportation. Often the indirect effects through such mediating factors as the ecology of the city were much more important than the direct relations themselves. In the short run one can identify many trips that do not get made because a letter or phone call takes care of the matter with greater ease. Just as the traveling salesman at the turn of the century could point to the trips he did not take, so any mother calling a doctor or manager calling the factory can easily identify trips saved. What one cannot report from common sense is how one's overall social world has been changed over time by the existence of a given communications system—and thus how the amount of transportation in society has been affected. These indirect effects are beyond our commonsense grasp.

The problem is paradigmatic of many in policy analysis; clear short-run effects of a policy can often be identified, but without being sure whether the long-run consequences are the same or indeed the opposite. Arms build-ups increase one's security in the short run, but critics argue that they make a more dangerous world. Wage increases help those who get them, but what if the resulting inflation damages the economy? What strategies are available for looking at such problems with something better than hopeless agnosticism?

One approach, which does not solve the problem, is to disregard the long run. But optimizing with respect to short-run considerations, of which we can be fairly sure, is to treat our ability to guess beyond that as zero, which is not accurate.

A better approach is to push the edges of short-run analysis toward some insights about middle-range consequences; the long run, after all, is but a series of short runs. Most serious research on the communications/transportation tradeoff is of this character. Let us briefly review three such types of studies.

Office traffic studies

In a number of countries, office activities have been studied to try to identify those that can be handled well by communications and those that require travel. Such studies are often done as a prelude to possible decentralization of offices—private or governmental—out of metropolitan centers. The question posed is how far the provision of good communications facilities can allow a move away from the center without causing a vast increase in travel.

Important work has been done by Bertil Thorngren in Sweden on the spatial distribution of contracts,[18] and by Chapanis and others[19] on the differences between communication with or without visual contact.

Alex Reid, who is responsible for some of the best studies, concludes that telephonic communication appears to be as effective or more so than face-to-face "for exchange of information among people who already know and trust each other, but that travel for face-to-face contact is important for creating relationships or for persuasion."[20] Thorngren found that most telephonic communication is with persons who are close enough for easy face-to-face contact and with whom there occasionally is such contact. In the absence of occasional face-to-face conversations the association attenuates; thus telecommunication and personal contact reinforce rather than substitute for one another.[21] From all this it follows that in organizations in which good working relations already exist and are being periodically reinforced by contact, improved telecommunications can save time and money. However, some meetings must continue if the organization is to remain viable and solve problems. If the sole goal were to reduce the amount of traveling, it would be better done by cutting all phone lines and letting organized interaction atrophy.

Econometric studies

Economic studies can measure the cross elasticities of demand between communications and transportation. The argument is sometimes made that the cost of communication is so low that decisions to communicate are made without much reference to their cost. In business, for example, where large deals are being handled, communication charges do not make one hesitate to facilitate matters by a telegram or long-distance call, which costs only a tiny fraction of the money at stake.

For the communications/transportation trade-off, the implication would seem to be that the prospects of substitution are not very good. If demand is inelastic, consumer use of communications in place of transportation will not be encouraged by price changes. However, econometric analyses of the elasticity of demand may, for two reasons, be too static for our purposes, making the conclusion dubious. In the first place, such analyses usually overlook the time lag that takes place in learning to use a new service. In the second place, econometric analyses of time series measure incremental changes and, unless specified with adequate structural equations, are likely to miss the prospect of very important tip points. There are two quite different mechanisms by which demand for any product grows as its price falls. One of these is the gradual greater attractiveness of the good purchased as its cost falls, which may be considered the trade-off between the purchased object and money itself. The other is the sometimes very sudden shift that may occur when there are two ways of doing something and one can choose the cheaper one. A study of the price elasticity of telegrams done before low-cost long-distance telephone calls came onto the scene would have shown some sensitivity of usage to price, but would have given no clue to the drastic change that occurred once it became cheaper to send long-distance messages by phone than by telegram. Under the changed circumstances the only continued major use of telegrams was for those of its characteristics for which a phone call was no substitute, such as having a written text for record.

My suspicion that there is likely to be a very elastic demand for telecommunications in the coming decades is predicated on the expectation that it may become possible to do certain things cheaply by telecommunication that are now being done in other ways. The most persuasive such case is the prospect that electronic mail and facsimile will displace manually carried first-class mail. If a cut in the cost of data processing, facsimile equipment, and transmissions brings the expense of an electronic message below that of a first-class letter, then no time series regression analysis of the years in which electronic mail was more expensive than first-class mail will give a good prediction of what will happen after that tip point is reached.

The prospects of a tip point might also be explored regarding information retrieval and library use, remote and local computing, telemarketing and shopping, or tele-education and schools. At least for large portions of some of these services it seems possible that the cost of teledelivery will fall below that of conventional delivery. These possibilities point to the importance of not taking the elasticity figure during a "normal" period as being predictive of these tip points.

There is good reason to believe that over the coming two or three decades a variety of different services will be delivered over a largely shared telecommunications plant. We have already discussed this convergence of modes. If information services, text, mail, and conversations increasingly use the same electronic network, then, conclusions reached by looking at the elasticity of demand for any one service may be misleading.

By now it should be clear that any simple statement that communications will either increase or decrease travel is not only empirically wrong but conceptually confused. As we can see from a simple model, the relationship can go either way, depending on the parameters of the situation, and indeed it can go both ways successively over a period of time.

The agenda for policy making will not wait for the social science research to be completed. Policy decisions rarely await ideal intellectual solutions. In general, it seems fair to anticipate that achieving an overall reduction in travel is not a very likely outcome of social policy, even in the face of severe constraints on energy resources. The goal of reducing total travel runs athwart of too many other social goals which depend upon increasing the range and cosmopolitan nature of human interaction.

On the other hand, reducing the incidence of unnecessary trips is a reasonable social goal, both for its own sake and for the sake of immediate energy conservation. If we find later that measures taken to produce such an immediate reduction in travel have second-order consequences that ultimately increase travel, these results would demonstrate marketplace evidence of the high value that people put on face-to-face interaction.

Regardless of ultimate trends, it would seem to be rational policy to provide improved communication facilities that offer a low-cost alternative to trips people would rather avoid. That can be socially useful regardless of whether the overall trend in society is toward increasing or decreasing travel. Marginal savings, and the associated, even if limited, direct impact of those savings on the size of the aggregate trend are reasonable goals for social policy. In short, making efficient use of available resources is a good idea regardless of how human beings in the marketplace choose to use the resulting bonus.

Megalopolis

What kind of settlements can we expect to see evolve in an age of two-way electronic terminals in every home? Some futurists have described twenty-first-century life in terms of a return to the countryside. It is a rather odd return to the countryside. The picture usually given is of a family indoors, looking out on a beautiful vista of mountains or sea, with no other human beings in view, and rarely emerging from the comfortable air-conditioned cocoon of their house. There they sit in front of a big screen two-way videophone and engage, by communication, in all the activities for which people now commute to work, to school, to shop, or to a doctor's office.

There is no need to censor such fictional fantasies for practicality or consistency. The rural fantasy of turning all the rest of mankind into shadows on the screen, manipulated like genies of the lamp, sits side by side in world's fair models with 100-story urban constellations with monorails like flying buttresses threading silently among the towers, and helicopters available for an occasional quick escape to the unspoiled countryside nearby.

As with all fantasies, there is much to be learned from them about human desires. People do crave an escape from overcrowded urban concrete deserts. They are fond of having their private piece of turf, but only if they can keep the comforts and human contacts for which they, or their parents or grandparents, exchanged rural for urban life. They tolerate commuting as a compromise, but resent the discipline of the relentless time clock.

Fantasies give us a clue of directions in which people may choose to use the devices coming to hand. There is reason to believe that the twenty-first-century fantasy of rural life reflects at least two uses that people will make of the increased powers of communications technologies. People are likely to conduct a good many activities from their homes that they now go to other places to do, and the megalopolitan pattern of settlement is likely to be even more dispersed than in the twentieth century, allowing more people to live in some contact with green things.

An increase in work at home would represent a reversal of a 300-year trend. Karl Marx, Max Weber, and virtually all analysts of modern industrial society have pointed to the separation of place of work from place of residence as a key feature. It began when the factory system in the sixteenth and seventeenth centuries replaced work by home craftsmen. To Marx, factories were part of the process of alienation of the worker from his means of production; the proletarian went to work at a plant that was none of his own. To Weber, it was a trend that applied not just to industrial production but also to intellectual production. He defined modern bureaucracy by a number of features, one of which was that the official owned none of his means of administration and did not work from his home, but went from home to office to work there with the organization's files. The same point could be made for education. The near universality of the modern-day school, to which the student comes, for a few hours, as if to an office, can be contrasted with many other patterns in the history of education before the nineteenth century: tutors came to the homes of the affluent; children went to the church, or parish house, or monastery, where men of the cloth who taught them lived; if they went to school it was probably to a boarding school where they went to live as well as to study.

The separation of place of work from place of residence occurred because it had many advantages. Some of these remain just as valid as ever, but some are sharply reduced when modern communications are available. The factory or office allows tight discipline by supervisors observing workers on the shop floor. To have the same degree of control of people working at home would require a video monitor over each worker, something that is technically possible but expensive to supervise, and not economical except for such special applications as security operations. But intermediate levels of control, far above what was available to the entrepreneur who rode on horseback every few weeks to the cottage of a weaver or a spinner, can be provided by telephone or by on-line data monitoring. For people who get paid by commission or by the piece, for whom minute-by-minute supervision is unimportant, telecommunications can easily provide all the information on the results of their work that an organization wants. Brokers, salesmen, editors, interviewers, and typists, for example, can be supervised quite adequately from a distance.

To some degree, then, people will find it convenient to work from their homes, and entrepreneurs will find it advantageous to have them do so. It will permit part-time work by mothers, use of specialists in remote locations, and reduction of labor unions' ability to organize employees. People can be hired on an ad hoc basis for peak loads, and a business can get started faster without setting up a facility, saving the employer some capital costs of plant and saving the worker the costs of commuting and of moving when changing jobs.

Collection of all the people involved in a project under a single factory roof had some advantages besides supervision and communication. It allowed for better handling of materials and equipment. Tools could be shared by more people and stand idle less of the time. Raw materials would have to be moved less. Storage and inventory could be done more efficiently. All those considerations continue to apply in an age of telecommunications. Where material handling remains important, as in a supermarket or factory, people are still likely to come to the job, rather than the job to the person. Still, if a major reason for assembling people vanishes, the balance between centrally located and dispersed operations will shift.

Noncommercial institutions may see many of the same reasons for changing their way of doing things as may business. Today students' work in school is closely supervised by the teacher, but homework is totally unsupervised. Telecommunications permits the adoption of intermediate patterns in which the child can get assistance while doing homework, check answers, and/or be set back on the track from time to time.

Physicians have taken to the telephone avidly; perhaps data communications networks can be equally useful to them. It is clearly possible to monitor a patient's temperature, pulse, electrocardiogram, and various other measures by telecommunications. Perhaps person-to-person visits between doctor and patient will become less frequent, in which case the mutual accessibility of their location will become less important. If medical telemetry comes into common operation, perhaps there will be a need for groups of medical practitioners on 24-hour duty monitoring the remote sensor data. Perhaps patients will look on a wider geographic basis for the appropriate specialty, rather than considering location in the neighborhood.

In all these ways, in an era of good telecommunication, location becomes a less imperious constraint on human activities. To that extent, people can, if they wish, choose to live in remote locations at a tolerable price. Few occupations or other activities permit total disregard of location. Even most writers and artists occasionally have to be where the publishers and collectors are; only a few can assume that customers will beat a path to their door. And most people need not only face-to-face business contact but also external stimulation. Hermits are rarely motivated and productive. Everything we know about morale and productivity stresses the importance of the small group. So constraints of location are only matters of degree. If a larger proportion of the population does some of their work at home and needs to come to a specific job location less often, that in and of itself will have an effect on the overall pattern of population settlement. More people will be willing, on the few days when necessary, to commute further; more people will spend longer periods in a rural second home; for more people their second home may be their city home; more people will invest in larger houses, because they will need work space in them.

Perhaps reduced constraint on location will operate even more on institutions and groups than on individuals. Consider computer programmers. To simply write code they can be located anywhere where there is a terminal, but to remain on top of the profession they must pick up ideas informally, such as over coffee and sandwiches with colleagues. They may be remote from where the physical computer is, but they will want to be in a community where work of their kind is going on. Or consider stock brokers. To simply fill a customer's order can be done anywhere over telephone and terminal. But to be engaged in serious financing, they must be where they can have lunch with bankers and businessmen. They must be where there is a community of fellow professionals in related activity.

These communities need not all be in downtown Paris, or London, or Manhattan. Increasingly, the trend in megalopolis seems to be toward a more complex division of labor. University towns are an example of dispersed communities. Each has to have a certain critical size and be part of an extended intellectual community by communication. But they are located often in the less densely settled parts of megalopolis. We have seen such communities emerge in other fields, too. Corporate headquarters congregated in New York's Westchester County. Electronic companies sprang up along Route 128 around Boston and on the peninsula south of San Francisco. They form specialized communities within a larger structure.

That is not an antipolitan trend; it is an extension of the more dispersed and more complex division of labor of megalopolis. There is no reason to believe that cities will disappear nor even their great downtowns. Telecommunications does not eliminate the human need to associate physically with other people; it complements such association.

Some writers have suggested that telecommunications will create communities without contiguity; that means that people will begin to form communities with others who share a common interest but whom they hardly ever see. Amateur radio hams and computer hackers are examples of such a community. No doubt that can and will sometimes happen. There is much evidence, however, that even in a society with a highly sophisticated communication system that will not be a dominant pattern. When Bertil Thorngren plotted the places to which people make phone calls, he found that most calls are within two miles, with the frequency falling off with distance. People who see each other develop relationships that lead them to call each other. Mothers phone to ask a neighbor to send the children home. People phone to make appointments or as a prelude to dropping in. Neighbors become friends and then phone to chat. Face-to-face contacts serve different psychological functions than phone calls. In a close relationship these two kinds of contact reinforce and support each other.

That is one reason why it is sheer fantasy to believe that telecommunications will cause people to choose lives of physical isolation. It is also unrealistic insofar as a large portion of human activity is not just exchange of information but involves joint action on physical objects. The proportion of persons who are processing information has risen enormously, and these people could (conveniently or inconveniently, cheaply or expensively) operate from a distance via telecommunications. But there is still the other half of the working population who are handling things. Assembly lines will not be abolished by telecommunications, though computers reduce the number of people working on them.

Truck drivers will still have to move products, even if they talk to each other by CB along the way. Haircuts are not going to be given by remote controlled automata—at least not in any period worth thinking about.

One can anticipate very large changes in the conduct of education, entertainment, science, the arts, government, finance, and administration in general. In all of those, the crucial activities can be carried out by remote communication. The effects of abundant and flexible electronic communication in other fields are less obvious, but they may be considerable. In manufacturing, for example, one may anticipate substantial further steps in the direction of individualization, exemplified by the emerging computer control of machines. Consider the possibility of an assembly line that does not produce items on speculation but according to individual orders. Now a customer either buys a mass-produced ready-made product or has a product custom made to order. Suppose there were a system by which a customer went into a shop to specify product type and material, and the specifications all were transmitted from a terminal to the factory, where the assembly line produced the item the same day. Suppose Detroit assembled cars after order under computer control in the same way. If something like that comes about, it would constitute a considerable impact of electronics and communication on the process of manufacture, but the basic process of physical production would still have much continuity with the way it is today. In housing, too, there is obvious continuity in the need for shelter, privacy, food, and sociability. Yet the design of housing might be substantially changed if families come to congregate in front of a six-foot television screen, and if both adults and children need work space by a terminal for their professional and school activities, presumably without interrupting one another.

Progress in communication is particularly important in a world in which there is increasing concern about scarcity of resources. There are no simple answers about how to keep improving the quality of life as energy becomes more expensive, or as the population/resource ratio becomes less favorable. However, it is clear that better use of information can increase the efficiency with which scarce resources are used. It is not clear what the optimal pattern of urban dispersion or concentration would be at different levels of energy cost and communication technology. There is a complicated interaction among those variables. Big cities are clearly more expensive places in which to do things than are smaller ones, but it is also more expensive to provide heat, food, and other goods to very sparsely settled populations. Somewhere in between there are optima to be found. In any case, compared with the present, it is certainly the view of virtually all planners that many activities ought to be dispersed out of the few overlarge cities that dominate most countries of the world.

In Japan, in France, indeed in most countries, planners want people to move out of the center to a more human and dispersed pattern of settlement. And most people when polled agree—that they would rather live in a place where they would have more room and less crowding if only certain kinds of jobs or services or educational opportunities were there. One untried way to achieve that would be to subsidize remote transmission of telecommunication services. In most countries in fact just the reverse is done. Long-distance telephone service is expensive and revenues from it subsidize cheap local phone service. We can speculate about what would happen if some government, as a policy, set a single uniform rate for phone calls (both local and long-distance) anywhere in the country, and set it at not much more than the present local call rate, perhaps if necessary, subsidizing the difference? It might markedly affect the pattern of settlement. Something like that was tried once before, forty years before the telephone, when the English reformer Rowland Hill proposed nationally uniform postal rates to achieve national integration and spur commerce. He pointed out that mail delivery was less distance sensitive than believed, since much of the cost was accounted for by local delivery, and only 1/10 of a penny per letter was chargeable to intercity transportation. The national penny-post was tried and became a great success. Very few countries today would permit the kind of penalization of remote areas that would occur if postal rates varied with distance.

Some economists object that such a subsidy to decentralization distorts resource allocations; but this is true for every subsidy. The argument for flat rates, in addition to their administrative simplicity, is that there are social costs in urban concentration. What or how extensive they are is an urban planning, not a communications, question. If in time such costs are found to be undesirable, an obvious way to offset them in part is by telecommunications pricing policy.

Communications, an Abundant Resource

There are no limits to the growth of ideas. If worry is justified about the scarcity of physical resources, all the more reason to improve the quality of life by use of resources that are not running out. Communications is one example.

If there is a reasonably well-working market, then those things that use scarce resources become more and more expensive; those that use less of them fall in price. Demand will respond to that financial incentive. The sacrifices forced by the rising prices of that which is scarce is offset by the ability to use more of that which is less so.

To have more energy, food, and transportation, one must use up some of the earth's scarce mantel and its elements. Communication, on the other hand, is a genuine addition to human welfare, but it does not necessarily use up physical resources. Granted, newspapers consume vast numbers of trees, but the information in the newspaper does not. It can be conveyed in resource-saving ways.

There is no obvious one-to-one relationship between the value of an idea and the cost of the medium that conveys it. A 50-million-dollar movie shown in large movie houses uses resources lavishly. Yet the same pleasure may arise from a school play. A thick Sunday paper with a magazine section is a large destroyer of trees and requires energy to dispose of it. A library book may produce as much enjoyment. In general the communicative activities in which people engage use much less by way of scarce resources than other major activities in their lives. That economy is measured by the small prices paid for using communications resources, be they broadcasts, periodicals, or conversation. Increasing the part communications plays in people's lives is a way of improving the quality of life with little drain on scarce resources.

Yet there are a few resources used by communication that one might worry about. Spectrum at once comes to mind, but we have already shown how that is an illusion due to the absence of prices. Four materials that might run short are copper, precious metals, plastics, and paper. But none of these, according to Michael Tyler and associates who studied material scarcities affecting British communications, poses a major threat to the expansion of tele-communications.[22]

The old AT&T used 15 percent of the copper consumed in the United States; in Britain, the corresponding number was between 5 and 10 percent of national copper consumption.[23] Copper prices, historically, have been notoriously volatile. The switch from iron to copper wires in the 1880s to accommodate the long-distance telephone system was itself made possible by a sharp decline in copper prices that had recently occurred with the opening up of new mines. The sharp rise in copper prices in the 1970s over-lapped (when prices doubled) with the oil crises of 1974. In the long run copper is probably one of those scarce resources which must be partly replaced if its price is not to rise a great deal. It can be replaced. Aluminum is almost as good a conductor and is often used when copper prices rise. It is considerably more abundant than copper. And then there are glass fibers, made from silicon, one of the earth's most abundant materials.

Paper is a more critical material. It is pervasive in our society and a potential bottleneck for the phone system—not just the press. Tyler's conclusion is that with proper forest management, paper supplies can keep up with demand in the coming decades. There have, however, been short-term paper shortages. Paper prices have risen greatly at times, becoming, along with labor resistance to automation, the central financial problem of the press. As computer-controlled editing, setting, and printing has been introduced, production costs have dropped, while paper costs have been rising, making paper an ever more important item of cost in publishing.

In the 1830s, when, under the impact of faster presses, the penny paper went on the newsstands in the United States, the paper used was made from rags. In 1843 the United States im-ported two million pounds of ragstuffs, and by 1850 twenty-one million; that figure doubled by 1857. Rag and paper prices soared; European newspapers went into the red; export restrictions were put on rags. In the crisis that ensued experiments were made in producing paper from all sorts of alternative organic materials. By 1870, newspapers all across the country were using wood pulp paper instead of rags.[24] The price of paper dropped from 30 cents a pound to 3.1 cents a pound at the end of the century. From 1870 to 1880, 5,429 newspaper titles started up in the United States. The penny newspaper spread to England, and then in the 1890s came the halfpenny commercial paper—enlarged, incorporating many features from magazines, and dependent on advertising, which followed from the marketing of paper-wrapped branded consumer goods. Cheap paper is essential to the press as we know it, and it seems likely that improved technology of production and forest agriculture will continue to provide it.

However, if contrary to expectations there should be a massive run-up in either copper or paper prices, it would have an effect on communications. There would be a painful period of adjustment; but it would not stop the development of communications. It would increase the incentive to distribute information electronically (to CRT displays) using optical fibers and microwave circuits in place of copper cables and wires or paper. Substitutions for a scarce resource are almost always possible. Early in the century some observers worried about the exhaustion of the supply of trees for telephone poles.[25] In time, underground ducts, microwave circuits, cement poles, and scientific forestry took care of that. Thus, it seems unlikely that electronic communication will run into serious problems of scarcity—outside of spectrum issues—that will force up its real cost to society and therefore reduce its use. As far as one can see, a long era of boom lies ahead.

If so, one moral is clear. As some resources get scarcer and more expensive it becomes increasingly important to improve life by means of those services such as knowledge, culture, and communication that are less wasteful of scarce resources. Communications will become an increasingly significant element not only in total GNP but even more importantly in those parts of it that are not under pressure of scarcity of natural resources and which can, therefore, help to quench the specter of the limits of growth.[26]

Chapter 12

Technology and Culture

Culture blossoms from mundane roots; the technologies of paint chemistry, metal construction, or physical recording have much to do with content in the fine arts and sciences. What happens to the technology of communication in the last decade of the twentieth century will resonate in the intellectual products of subsequent decades. On the nature of these consequences we can only speculate. Changes in thought occur slowly. Early in a lifetime a person acquires most values, skills, and aspirations that shape his thinking for the years to come. The economic and institutional impact of computers, video cassette recorders, photocopiers, or facsimile machines can be assessed fairly early on, at least in the small picture; to consider what the expansion of those initially tiny but visible early tendencies may mean to the larger culture is a rather disciplined kind of speculation. But how thinking will change among still malleable young people who grow up taking the new technologies for granted is not seen that quickly.

Culture is a fragile flower. Under some social circumstances it blooms magnificently; under others it withers and dies. The Technische Hochschule in Budapest, in one decade early in this century, numbered among its students John von Neumann, Theodor von Kármán, and Edward Teller. Out of hundreds of cities in the world, in one small city, ancient Athens, the bases of Western science, philosophy, and literature were established. Across the Mediterranean from Athens, a small tribe living in a land the size of Rhode Island laid the foundations of several great world religions. In Paris, during half a century, virtually all the titans who created modern art could be found living and working within a few blocks of one another.

Explanation of such explosions of intellect has eluded psychology and the social sciences. There are many contributing causes, and most need to fall into place. Some of the conditions are themselves intellectual, such as the seminal ideas of a genius; some are highly mundane, such as the existence of a cheap and permanent material on which to record. Repression, poverty, isolation—each alone is enough to snuff out an intellectual explosion. There is, therefore, no way to forecast whether, or where, the new technologies of electronic information processing and communication will spark an efflorescence in science, culture, and the arts. But we can consider the intellectual styles that new technologies may make possible, without presuming to guess where, when, and whether they may mesh together into a great renaissance.

Interaction and Diversity

The communications technologies evolving today are less frozen into the uniform output of a mass medium than communications technologies of the past; they hold the promise of being shaped to the needs of small audiences or particular individual users. The reception of mass communication is passive to a large degree; the new communications technologies are adapted to more active information seeking.

A standard criticism of television is the passivity of using it. It comes right to the home where one is already located. There are so few channels in most of the world that the viewer has little choosing to do. Where choice exists, the individual viewer usually has to compromise with the taste of the whole family; only the affluent can afford sets for individuals. Watching television takes little skill; a peasant on the first day has enough visual literacy to make something out of it. It is seductive enough to invite surrender to it. And it is a one-way medium.

In contrast with print, one feature of limited television is that the viewer cannot jump around to only those items that are interesting and congenial to him. A reader skims newspaper headlines and stops at stories that are meaningful to him. A viewer, if he watches a newscast at all, sees in sequence the dozen or so items the producers put on. Some of what passes before his eyes he may disagree with; some pictures are of people he dislikes. Research on newsreading shows that readers tend to skip such items; they evade incongenial material. But on television they have less choice.

Michael Robinson tested people's reaction to a TV documentary biased against their predisposition. Disturbing material did not persuade, nor was it rejected; it was psychologically turned off.[1] The result was not a better informed citizen, or a persuaded or argumentative one, but an uninterested citizen. Television as presently structured is thus, according to the evidence, not a medium adapted to mobilizing and involving the public except when there is already a simple congruence of its message with their predisposition. It is rather a medium for passing time, for being amused, and for taking it easy. But now the pendulum of technology begins to swing away from providing such passive media toward creating ones that require much more vigorous interaction by the audience. What is that likely to mean?

Access programming

A set of issues has already surfaced in the debate about "access" to the studios of cable television. Small off-beat groups have seen in cable TV an opportunity to make their own videotapes and broadcast them. They almost never succeed in winning a substantial audience. Critics of access programming cite that fact to brush them off, as though the test of the merit of their activity was the same rating game that is appropriate to scarce over-the-air channels that have to serve everyone. The cost of access programming on cable is low enough to justify its use as an experience for those who make the programs and their handful of friends who watch.

What the access groups now do with the time they are given may be a poor clue to what would be done by wider circles if cable or a successor carrier like optical fibers becomes a major medium and channels become liberally available. The first groups to seize the opportunity for access are in general young and alienated from the mainstream recreations of the society. To a large extent the thrill that present access programmers get is from showing faces that are much like themselves to the world that is assumed to be out there watching. Like cinema verité, it is telling the world to look and see what is around them. While that kind of self-expression may well survive, and is indeed a useful function for access channels, there is every reason to speculate that if the making of video becomes a widespread skill in schools and elsewhere in society, there will be much increase in purposive use of it by groups with more pragmatic goals. These goals could be in education, or publicity, or community action. One would also surmise that there would be a growth in those who use the art form with skill as a kind of aesthetic expression. All of that remains to be seen. But in any case, video would be a very different thing if substantial numbers of people were engaged in it as an expressive medium, and there were channels enough so that they could each mobilize their own clienteles.

There is no reason to assume that with greater diversity video would become, on the average, a higher quality medium than it is today. It is a standard illusion to believe that new options will somehow advance culture and taste. Those who forecast radio broadcasting and TV, for example, generally foresaw them as having high moral quality and raising public taste.[2]

Some media may do that, but the multiplication of broadcast channels, wherever it happens, seems to have a different effect. Indeed, the argument is often made, for example in the Annan Commission's report on British television, that quality arises from monopoly, and that fractionating the audience produces a downward competition for audience and thus a lowering of standards.[3] The basic argument consists of several explicit or implicit propositions:

1. When offered a choice between programs, most people will choose the one at the lower standard.[4]
2. If media have to compete for audience they will, therefore, engage in competitive lowering of standards.
3. High-quality production is expensive, does not pay for itself in expanding audience, and therefore does not pay for itself in drawing advertising.
4. High-quality programming will, therefore, be done more often out of monopoly profits than by organizations which have to compete economically.
5. Competitive channels seeking the same audience and the same funding will therefore converge in presenting very parallel programs aimed at mass taste; monopoly will offer more diversity.

The American situation is usually pointed to (as for example in the Annan report) as empirical evidence for these propositions. American radio illustrates that very standard uniform programming can result from many players all competing for shares of the same pot. Even the American TV networks used the radio example as an argument against adding a fourth commercial network, arguing (unlike pure monopoly advocates) that three networks give diversity, but more would fractionate the audience in an economically disastrous way and bring quality down. The Annan report used the same argument against even the three U.S. networks and concluded that in Great Britain it would be better if competitors draw their revenue from different sources. (The BBC draws its revenues from license fees, while Independent Television is financed by advertising.)

The whole argument, however, in its simple form is much of a debater's brief, though there is something to it. For any validity, proposition 1 has to be restated. The distribution of tastes in the public spans a wide range; competition for the audiences will drive stations toward the modal level—not necessarily down. The reason that intellectuals and elite policy makers see this as downward pressure is that it is inconceivable to them that programming offered on a controlled medium might be below the mean of popular taste. Their goal is to protect programming above that mean.

Proposition 5 is the most seriously defective. In economics there is a large literature on monopolistic competition and market segmentation. When there are only two or three competitors there is much to the argument as presented. Then each competitor strives for its half or third of the audience, which means aiming at the central tendency of the market. If there are ten or twenty or thirty competitors, however, each can hope to win on the average only a small percentage, and the best strategy becomes that of finding some special-interest group and winning their loyalty for one's own differentiated product. Diversity under those circumstances comes from numbers of suppliers.[5]

When there is diversity, some stations will aim at lower and some at higher quality. It matters little where the average quality level falls; the multiplication of channels means that there will be some channels cultivating high taste.

A problem that remains is for these more specialized stations to find the economic resources to do programming of high quality. That brings us to the issue of pay-television. If the only possible source of revenue is grants or advertising, or a license fee, then indeed it is hard to find such revenue. That is analogous to the system by which books used to be priced in the Soviet Union. The pricing had nothing to do with costs of production or expected sales; the formula set the price of a book at a certain number of kopeks per page. Clearly no publisher had an incentive to produce a high-quality small edition for he could not charge more for such a book. Where a genuine book market exists there are books of all qualities and all sorts of prices. So it would be, too, with pay-television. The reward for doing high-quality programming for a small audience has to be that people who want it enough pay accordingly.

As long as the system does not allow people to vote the intensity of their sentiments by what they are willing to pay, it is true that minority tastes are penalized and that the only way to assure that some elite taste will get served is to set up a monopoly covering some portion of the revenue (as the Annan Commission recommended) and leave it to the benevolence of the monopolist to occasionally cater to that privileged taste.

Technologically, means are becoming available to allow a wide variety of video programs, including some highly sophisticated cultural material. Abundant cable channels can be made available at fairly low costs; electronic feedback on the systems allows by-the-program billing or simplified subscription to special channels.

Pay-cable TV certainly could operate under the same market mechanisms that exists for print; that is what the Whitehead Report in 1974 advocated for the United States.[6] The report, unfortunately now largely forgotten, was a last desperate effort by the White House's Office of Telecommunications Policy to commit the United States government to force cable TV into the print model under the First Amendment. Such a system would presumably lead to mass entertainment and news channels, drawing much of the audience, attracting the bulk of advertisers, and thus able to provide the most attractive programming. Side by side with these mass appeal channels would be specialized offerings, some appealing to low tastes and some to high. What has happened in print is our best forecast of what might happen in a pluralistic, competitive, open-entry video system.[7]

But if print culture is one guide to what video culture can become, it is at best a limited guide, for the role of print is undergoing change, too. The change is toward interactivity; video will share in that development, too.

Interactivity media

The written word is canonical. Deemed sacred in the Judeo-Christian tradition, even in secular society it is regarded as more serious, more definitive, requiring more careful consideration than oral statement. Oral communication is interactive. An oral statement is treated as putting a thought on the agenda; it is a trial balloon to which another is expected to respond, and out of the discussion may come some wisdom.[8]

A device using electronic logic exudes in this respect more oral than written communication. It too can be interactive. A culture that substitutes interactive text for written hard copy may undo what was earlier done when hard copy replaced dialog, and in the process produce a new synthesis. Written texts, and particularly printed texts, lend themselves to exact repetition and so to caring about it. Among the priestly orders of the cult of getting the word right are typesetters, proofreaders, copy editors, and lawyers. Versification had served the same purpose for oral narratives; one could learn and check the exact wording of a prayer or epic with the aid of its meter. But in most oral communication it is enough if the spirit is recaptured in one's own reformulation.

Writing establishes hierarchic relations between the ones with the talent or authority to originate texts and the others who read them. Readers in turn are above illiterates to whom the text must be explained. Printing made society more pluralistic by increasing the writers; new pamphleteers, from Puritan parsons to Walter Lippmann and James Reston, gained authority from print. A clear-cut hierarchy separates those whose every word will command millions of readers, those who struggle to get published, and those who would never think of themselves as able to get a word in print. Some of the latter are perfectly willing to express their opinions orally, or even shout them loudly and walk on a picket line; but writing they leave to others.

Excellence may be rewarded by the hierarchic and repetitive character of the culture of print. The hundred great books, or however many there may be, at the core of civilization get preserved and studied by generation after generation of students. In the days before printing, important writings were the only ones worth the time and cost of copying by hand. Since printing, much junk gets reproduced, but classics survive in hundreds of editions and millions of copies.

A process of selection has been modeled by William McPhee in a survival theory of culture.[9] The limited amount of truly great material that a society produces is the wheat; along with it comes much chaff. The average level of a culture can be marginally improved by reducing the amount of wheat accidentally discarded, but to a much greater extent the average level is determined by the proportion of chaff cleared away. Network television necessarily resorts to an enormous amount of chaff; there is not enough genius around to produce first-rate material for every hour. An anthology of poetry, on the other hand, sweeps aside all the millions of bad poems that have been written and presents just a hundred or so of the best. But that quality is bought by enormous repetition; any two anthologies of a culture's great poetry will contain many of the same poems. Extraordinary material comes to the top only by selection and repetition of great works. Freshly produced oral material may seem pleasant enough when people entertain themselves with it, but a tape recording of a delightful party is abysmal when preserved.

Formal education was a product of writing and shares its hierarchy, repetitiveness, and often excellence. Classic education was the close study of great texts. Before print made copies abundant, students memorized texts. With print, memorizing was limited to a few poems and ritual statements, but the student remained a passive receptor of the written thought of the ages. In the early twentieth century, progressive educators like Maria Montessori and John Dewey rebelled. The roots of the change were partly in science, which requires physical interaction with equipment in the lab. A conversation with another human is at one pole of a continuum; a mass medium is at the other. Other devices lie along the continuum; while not conversation, each intermediate medium shares a part of the satisfaction of being interactive.

A musician interacts with his instrument; in a sense he talks to his violin, and it responds—or sometimes fails to do so. Even someone picking records off his shelf is more active than someone listening to the radio. Photocopying of the pages you want to keep can be a substitute for reading them carefully, but like note taking it does require an interruption of a somnolent flow of reading in order to decide what to keep. Working with a camera is, of course, active. But compared with these, a quantum jump toward interactivity is represented by on-line computing in its various forms.

Computer-aided instruction is one form. Information retrieval is another. Computer simulation games are a third. Whatever the form, the experience among people who have tried these is excitement, and often also intense frustration. The experience of children is perhaps most positive. Patrick Suppes of Stanford, and Seymour Papert at MIT, among others, have shown that children at the keyboard find learning fun and exciting.[10] There is some as yet unanalyzed magic in making the great machine come back and respond to you, of finding that it resists your demands but finally mastering it.

Adults have the same reactions but are more reluctant to learn to adapt, as, for example, by learning how to type. With present terminals, the necessity for typing as input is one of the obstacles to growth in use. Executive types and uneducated persons alike are threatened by being sat at a machine that secretaries handle so fluently but on which they have to hunt and peck and correct innumerable mistakes.

An equal obstacle to adoption is recognition by users of how blunt and cumbersome many early computerized information-handling systems are. There is a literal-mindedness about computers that requires that one feed into its tabula rasa every needed distinction that the user has in the back of his mind from a lifetime of experience. If a sophisticated novice wants to ask a subtle question, the tutoring programmer has to explain to him how to define his problem with a host of "ands" and "ors"; by the time the obvious is made explicit the whole thing has lost interest. This is illustrated by the difficulty of writing programs for computer-aided instruction.

But these barriers to rapid adoption of computerized interactive systems are the problems of an early stage of a new art which has a long way to go. Gradually, as systems grow and improve and fall in cost, the power and flexibility and adaptation to the individual user of interactive computer systems may well conquer a new generation of information seekers. They will no more understand how we tolerated the tediousness of a manual search through files than we can understand how people put up with the tediousness of copying reports by hand before the typewriter and the photocopier.

A question that we can only pose is how far interactivity may become a habit. Will children who grow up on terminals find passive watching of TV less satisfactory than do today's children? Will people who get used to using information retrieval systems that allow them to ask one question after another in "20 Questions" fashion be less patient to read someone else's essay or book? How far will intellectual life attract a different and more activist personality type when more of the operations become active? To answer such questions with confidence requires prophetic vision, for the answers have not begun to emerge yet.

The Future of the Book

It is common at this stage in discussions of new systems of information handling to suggest that perhaps the book as a form of communication is about to die—or in a similar vein to pronounce an obituary for the newspaper or magazine.[11] I do not propose to do so. Change is not death, though it is painful, too.

Books have changed a great deal in the past and they will change again in the future. In the Middle Ages manuscripts on parchment were expensive and rare. They were too expensive to use casually; most were of important classic texts, so a main activity of scholars was to interpret those texts. With printing it became much cheaper to produce books, and so a life purpose of scholars came to be to write books. The original book-length treatise became the symbol of intellectual achievement. That was true not only for novelists like Cervantes, or philosophers like Rousseau, but even for natural scientists like Galileo or Newton.

But that was before the mass media revolution. While more full-length books are appearing than before, they have increased much less than more fugitive material. Magazine stories and television serials are the hallmark of the twentieth century. Newspaper columns and editorials are our main forum for ideological discussion. And scientists publish their important discoveries in journal articles which will be out of date and forgotten in a decade.

In the nineteenth and twentieth centuries magazines and newspapers were charged with threatening the death of serious writing in books. "The habit of desultory miscellaneous reading has, alas! taken firm root," wrote a critic in 1906.[12] American book publishers in 1912 opposed second-class mailing rates, and insisted that each class of service should pay its own way, for they saw the government subsidizing their competitors, the magazines.

Then came the challenge of movies, radio, and television, each of which in turn was deplored as a threat to the book. Said one of Edison's associates:

To the final development of the kinetographic stage, than which no more powerful factor for good exists, no limitation can possibly be affixed . . . Not only our own resources but those of the entire world will be at our command, nay, we may even anticipate the time when sociable relations will be established between ourselves and the planetary system, and when the latest doings in Mars, Saturn and Venus will be recorded by enterprising kinetographic reporters.[13]

One critic in 1929 lamented, "A library will consist of a store of talkie-films,"[14] while Vachel Lindsay in 1915 exulted, "There will be available . . . collections of films equivalent to the *Standard Dictionary* and the *Encyclopaedia Britannica* . . . Photoplay libraries are inevitable."[15]

What effect would these new media of movies, radio, and television have on print? Their rise in the 1930s stimulated literature on the question of how a new medium affects the audience for an old one. The most important treatment was Paul Lazarsfeld's *Radio and the Printed Page*,[16] but the question was addressed by many others, too.[17] The particular question with which Lazarsfeld started was whether radio news bulletins increased or decreased the readership of newspapers. It could go either way; listeners having heard the news could feel that they had no need to read it; or, their interest having been piqued, they could become more avid readers of the fuller stories in the press. The same question could be asked about the impact of a movie version on the sale of a book, or the effect of the three daily hours given to television on the time spent reading. The answers, as we have seen, are varied depending on the completeness of the equivalence. Television cut a bit into reading time, but its main impact on publishing was to compete with magazines and newspapers for advertising. Books survive, writing survives; Marshall McLuhan to the contrary, pictures are not their functional equivalent, nor is an ephemeral news bulletin the functional equivalent of a newspaper. Books survive, yet in growth books have been outpaced by more fugitive devices that give the story for the day and are not carefully shelved to be used again tomorrow.

Now new technologies based on computers and electronic

transmission raise the same questions about the future of the book all over again. The outcome will not necessarily be exactly the same, though the experience has much in common. An individual computer reply flashed up on a CRT is even more ephemeral than a news story in print or one on the air; but the power of computer information handling is a synthesis between the fleeting character of the conversational reply and the permanence of print. All the ingredients that produced the computer output are (unless programmed to be purged) somewhere in store, able to be retrieved again on demand. Vachel Lindsay's notion that visual images would be much used in library fashion did not come to fruition largely because no one knows how to look up a picture easily; we do not know how to index or browse through films easily. What is in nontextual media is powerful for its moment but hard to recapture. Computer interaction is fugitive, also adapted to the moment, but not necessarily hard to recapture. So the area of its competition with print and with books is considerable.

Consider the census volumes. The many expensive volumes of the census collection on the shelves of a library are a very inadequate substitute for an on-line information retrieval system. The volumes are a report on some 200 million questionnaires. Someone at the Bureau of the Census had to decide what tables to run, including usually no more than five variables. Since the 1970 census, however, we have been moving toward a better solution. A sample of the population, say one in a thousand, is put on tape, minus, of course, information on address or other unique characteristics that would make a respondent indentifiable. That tape can be loaded on one's computer or accessed through a network to allow any researcher to make precisely the table that is wanted. While the demand by researchers for ever more voluminous and detailed tables grows, in the end the printed shelf of census volumes will probably become smaller. Those basic tables to which people refer often may be printed in books; for the rest, the data will be stored somewhere, accessible to anyone on-line.

That conclusion, as we have noted elsewhere, does not imply the desirability of building only a few giant central archives to which everyone everywhere will have remote access. On the contrary, it implies the end of such primary depositories and the retention of much data in its normal place of origination. There is a limit to what can be produced in hard copy for library deposit.

Expensive activity in archiving can be reduced if networking permits data to be accessed in its natural habitat, where it accumulates in the process of daily work. As we noted earlier for manufacturing lines in general, the optimal balance between publishing in advance and publishing on-demand is shifted by computerization in the direction of the latter.

At this point the reader's reaction might properly be that we have used as an example the most obvious case—computer data bases of a factual kind, like the census. Does it have anything to do with the great bulk of published material: novels, essays, tracts, poetry, textbooks, cookbooks, children's books? It does.

Reference compilations like the census, the consolidated airline guide, bibliographies, and directories are indeed the first and most obvious publications to move from hard copy to electronic memories, but to a degree it may well happen with a variety of other books, too. We can leave it to science fiction writers to contemplate a person sitting in front of a terminal to read a novel. We have no basis for speculating about that. What we must contemplate is the fact that for virtually everything in writing, it is likely to become cheaper to store it and send it electronically than to send it on sheets of paper.

For essays, history books, textbooks, letters between friends, business records, and all other writings except where aesthetic considerations control (as in scented love letters) or where portability is important, as with paperback novels, it will become hard at some point in the future to justify the economic expense of keeping records on paper. It costs money to keep files in filing cabinets or on library shelves.

It is already cheaper today to store a document or letter in mass computer storage than to keep it as a piece of paper. The reason that paper still usually wins out is the high costs of putting the document into computer storage in the first place and of getting it out of computer storage when you want it. These costs tip the balance in favor of traditional files, but that may change in the next decade or two. If a record has been entered into a word processor or personal computer (and already much of all that appears in print goes through a computer at some point), electronic storage beats paper files.

Now, that is likely to have an impact on all serious intellectual activities, not just on reference handbooks. One effect is to obliterate the line between manuscript and final copy. Today an author produces a draft, circulates a few copies perhaps for comments, and then sends it to the publisher, who turns out the canonical copy that remains fixed at least until a second edition a few years later.

Increasingly today there is a subculture that operates quite differently. It is the subculture of authors who use computer networks. Such authors type a manuscript into computer memory from a terminal and give access permission to colleagues to read it. When they read it these colleagues are likely to type their comments into computer memory as well, and to make them available to others, too—or not. The author can make changes as desired. Thus, the manuscript as stored in the computer may change daily. There may never be a canonical version. What the computer will print out on any given day will be the state of the text on that day. Authorship becomes a collective endeavor, and ownership rights (such as copyright) become a murky affair.

Also, different persons may copy the text and modify it to suit themselves. For example, suppose a teacher uses a certain textbook but wants to add some specific examples. If the text is stored in the teacher's computer memory, the teacher can make the changes, after which the book will exist in two versions. As each student makes his own marginal notations, the book comes to exist in hundreds of versions. Ironically, we are back in the situation of the handwritten manuscript with all its variations. What we may expect increasingly to find is constantly changing printed texts in infinitely varied forms, originating all over the world, and accessible from all over the world.

That notion applies not just to reference books but to all kinds of books. Someone writes a poem. In the old days of the oral tradition, each bard varied it, improving it or making it more topical, or gratifying his own or a sponsor's ego. In the era of the printed book, the text remained inviolate. In the time of editable computer files, it becomes easy again for an imaginative reader to express his own creativity in emendations. Whether a third person sees that as an improvement or not, there is much to be said for it as a more involved way of enjoying poetry. It is not my purpose to say this is all good or all bad. If it were all bad it would not happen, for people will reject those devices that serve no useful purpose for them. If it were all good, our present books and mass media would simply disappear—which I do not think they will. All experience with changing media point to the continuity of old forms. As new forms come in, older forms persist—the continued use of the book in the age of mass media, the continued use of print in the age of broadcasting. Books are cheap because paper is an economical storage medium even if more costly than it once was. Books are easy to browse through. They can be read on a bus, or at the beach, or, for that matter, in a bathtub.

Progress in the technology of communication promotes as well as competes with book publishing. The same technologies that we have been discussing make it possible to produce books much more cheaply than in the past—or, to be more accurate, to partly offset the rising costs of paper and labor. Computer composition and editing of the original manuscript in many instances permit the publisher to produce photoready copy with no typesetting or retyping at all. Photoreproduction allows small press runs, reducing risks and saving on inventory. For high-quality production, lithography and color allow an elegance of product in even inexpensive books that was inconceivable half a century ago. A modern fashion or trade magazine is a work of art worthy of much more elevated content. Thus there is no reason to assume that technology will fail to keep physical books a marketable product of value, just as much as it will impose change upon them.

What I am forecasting is the growth of different options and with them a change in the balance of media chosen for varying uses. The choice between continued use of traditional print publications with all their uniformity and rigidity on the one hand and expanding use of the fluid, individualized output of computer-information handling and on-demand publishing on the other depend on the balance between two sets of considerations. It depends upon the balance between the need for work-saving guidance on what to read or listen to on the one hand and the desire for custom tailoring of information on the other.

For the scholar or the specialist working in his own field, dynamic information handling is ideal. A scholar or business specialist would be delighted each morning to have a newspaper devised for him which included in full the stories dealing with the topics of his professional concern. What is cut out by such a device is the newspaper editor. But editors have their uses! Unless someone is very purposeful in information seeking (as in professional research) then he is usually happy to have someone else tell him what is important and what he should be interested in. Every successful mass medium is such a channeling device.

In principle, one can read any one of the million or more books in the library, and when one is doing professional work one may indeed burrow through books that no one else has taken out for 50 years. But an ordinary recreational reader tends to read books that are reviewed or advertised and that one hears others talking about. The great majority of books read in any given year are the few current best-sellers. Most of them will be soon forgotten. For most readers the rest of the books in the library might as well not exist. The publishing industry could not exist without such guidance mechanisms for readers as reviews and planned merchandising.

So it is as well with movies. Where are the hits of yesteryear? The blue-sky notion that in the future one will be able to dial up any television show one wants out of a library of everything ever produced misses the point. Much of the public may not want such individual responsibility. Yet the physical capability for providing such a service with dynamic, changeable, individualized material will be attractive to specialists and aficionados. An enormous expansion of the available knowledge base is occurring, and for some people, at least, there is the option of working on it with constantly changing and individual variants forming an elaborate mosaic of records.

Copyright

The flux of infinitely varying electrostatic and computer copies wreaks havoc with copyright, of course. The concept of copyright is rooted in the technology of print. The recognition of a copyright and the practice of paying royalties emerged with the printing press.[18] When numerous copies were reproduced in one place, it became easy to identify the source of the copies and how many had been made. That plant was the practical place to apply any control or fiscal accounting. Indeed, the practice of copyright in Britain, though not the world, began in 1557 when Mary Tudor, in an effort to stop seditious and heretical books, limited the right of printing to members of the Stationers' Company and gave the

Company the right to search for and seize anything printed contrary to statute or proclamation. Eight years later the Company under that power created a system of copyright for their members.[19] In 1709 the first copyright act for authors was passed.

While the concept of copyright was not directly tied to the printing press, for those modes of reproduction where easy means of control did not exist—conversation, speeches, songs (until quite recently)—copyright restrictions were not applied. Copyright was a specific adaptation to a specific technology, and to the problems and opportunities it created.

The law recognized that. The landmark case in the United States was *White Smith v. Apollo*. It denied protection to piano rolls or sound recordings because they were not "writings" in tangible form readable by a human being.[20]

That common law concept of copyright excluded from protection many new technologies of communication since 1908. But the motion picture industry, the recording industry, and more recently the broadcasting industry have persuaded the Congress to extend various protections to them, since courts were not willing to do so. For the earlier new technologies—movies and phonograph records—the logic of the extension was reasonable. Like books, they were physical objects produced in multiple copies in some sort of production plant. The same system as had been applied to printing some centuries earlier was basically workable. However, with the arrival of radio and electronic reproduction, and now photocopy reproduction, the concept becomes inappropriate. There is no easy way to keep tabs on the numerous reproductions in somewhat variable form that can take place in innumerable locations with these new technologies. The analogy is to word-of-mouth communications in the eighteenth century, not to the print shop of that period.

Nonetheless, information and publishing industries whose welfare and survival depends on finding some way to charge for their information processing services have latched on to copyright protection under statute law, and are trying to get the courts or the Congress to extend copyright protection to computerized data, photocopies, and telereproduction. Though recognizing that in those technologies the existent copyright law is basically unenforceable, they nonetheless grab on to whatever frail reed it may provide rather than turn to the even frailer reed of trying to invent, and to get into legislation, some entirely new as yet undevised system for rewarding the creators of information.

After several years of legislative struggle, the U.S. Congress passed a new copyright law in 1976,[21] which was designed to solve all the new problems of copyright for cable television, photocopying, and computers. It has solved few if any of them. It has provided a set of rules that will be cited for some years as a norm, but presumably violated as often as obeyed. The new law extends copyright to 50 years beyond the author's death, thus bringing American practice in line with that of the rest of the world, but at the same time extending an unworkable system to an ever larger number of works no one cares about.

The law could impose serious burdens on the development of electronic information handling. Unenforceable laws, like the Eighteenth Amendment, may not prevent individuals from doing what they want to do, under the cover of privacy and corruption, but they do prevent substantial responsible institutions such as corporations and universities from addressing themselves to meeting the public demand effectively.

How inappropriate the concept of copyright is to computer communications becomes evident as we examine how the law has to squirm to deal with the simplest problems. The primary purpose of the copyright law, as declared in numerous cases, is the promotion of science and the arts.[22] This goal is generally superior to the additionally stated purpose of protecting the tangible intellectual labor of the author of a scientific or artistic work. For the work to merit protection, it must possess some "creative originality." It must reflect the author's own skill, labor, and judgment and must be more than an industrious collection of previously known material already within the public domain.[23] While publication is all that is required of material an author desires to protect,[24] registration with the Office of Copyright perfects the right of the author to bring an action for infringement. In sum, publication protects, while registration perfects, the copyright. Those works which are unpublished remain unprotected by either American statute or case law. Under Common Law, however, unpublished material is protected. This becomes an important consideration in the computer field when one notes that some original work may be found in the computer which is the forerunner of published information. Such pre-published material is not now granted protection based on statutory copyright. That portion of a news story, for example, which is entered and stored in a newsroom computer, although excluded from the final article, is fully the original work of the author-reporter. Yet, while it remains in unpublished or pre-published form, it remains unprotected by domestic copyright statute. Consequently, there is some risk to authors that their computer-posted background material may be used by those with access to the stored information. For authors whose works are based on computer-stored information and artful programming, there is a danger that the "raw material" of intellectual labor may be accessed and used without violating copyright.

The *Apollo* case held that infringing copies (and therefore protectable material) had to appear as visible intelligible notations that were humanly, not machine, readable. It has traditionally been understood that that which is part of a machine process is more likely to be within the realm of patents and, therefore, ought to gain its protection from Patent Law provisions.[25] In the modern period of computer-information storage systems, the dilemma is posed that the computer recording of information may be neither an infringing copy nor a protectable entity. The current provisions of the Copyright Office accept programs, regardless of their mechanical character, for registration and protection as a "writing." This is based on the notion that, even though the recorded information appears in a form difficult for the human being to read, it does record material which, when processed, gives a readable print-out as the end product.

If the material is protected, can it be argued that a program which effectively requests an analysis of stored information with a final product in written form (the print-out) is in fact a computer-authored or co-authored electronic publication? The idea that a machine is capable of intellectual labor is beyond the scope of the Copyright Act. If the computer is not the author of something automatically produced, then who is?

Can a computer infringe copyright? The printed output of recorded copyright material is likely to be a statutory violation of the Copyright Act which vests the exclusive right "to print, reprint, publish, copy and vend the . . . work."[26]

In short, the process of computer communication entails processing of texts that are partly controlled by people and partly automatic. They are happening all through the system. Some of the text is never visible but is only stored electronically; some is flashed briefly on a terminal display; some is printed out in hard copy. What started as one text varies and changes by degrees to other things. The receivers may be individuals and clearly identified, or they may be passers-by with access but whose access is never recorded; the passer-by may only look, as a reader browsing through a book, or he may make an automatic copy; sometimes the program will record that, sometimes it will not.

To try to apply the concept of copyright to all these stages and actors would require a most elaborate set of regulations. It has none of the simplicity of checking what copies rolled off a printing press. Good intentions about what one would like can be defined. One would like to compensate an author if a computer terminal is used as a printing press to run off numerous copies of a valuable text. One would like not to impose any control as someone works at a terminal in the role of a reader and checks back and forth through various files. The boundary, however, is impossible to draw. In the new technology of interactive computing, the reader, the writer, the bookseller, and the printer have become one. In the old technology of printing one could have a right to free press for the reader and the writer but try to enforce copyright on the printer and the bookseller. That distinction will no longer work, any more than it would ever have worked in the past on conversation.

Those whose livelihood is at stake in copyright do not like that kind of comment. They contend that creative work must be compensated. Indeed it must. Publishers may point the finger in accusation and charge that one is taking bread out of the mouths of struggling writers. But the system must be practical to work. On highly charged subjects there is an impulse to insist that those who make a negative comment must have a panacea to offer instead. If one says prisons do not cure criminals, the rejoinder is apt to be, "Do you want to let them out to kill people?" One does not necessarily want that at all, but it may still be true that prisons do not cure criminals. Likewise, one can say that in an era of infinitely varied, automated text manipulation there is no reasonable way to count copies and charge royalties on them.

That is the situation now emerging. It may be very unfair to authors. It may have a profoundly negative effect on some aspects of culture, and in any case, whether positive or negative, it may change things considerably. If it becomes more difficult for authors and artists to be paid by a royalty scheme, more of them will seek salaried bases from which to work. Some may try to get paid by personal appearances or other auxiliaries to fame. Or the highly illustrated, well-bound book may acquire a special marketing significance if the mere words of the text are hard to protect. Or one may try to sell subscriptions to a continuing service, with the customer knowing that he will be a first recipient.

These are the kinds of considerations one must think about in speculating about the consequences for culture of a world where the royalty-carrying unit copy is no longer easy to protect in many of the domains where it has been dominant. While Congress tries to hold the fort, it is clear that with photocopiers and computers, copyright is an anachronism. Like many other unenforceable laws that we keep on the statute books from the past, this one may be with us for some time to come, but with less and less effect.

Some Recapitulations

A few additional points that I have dealt with extensively elsewhere in this book have special implications for culture. For example, no doubt the erosion of distance has a tendency to make cultures around the world more cosmopolitan, whether or not that also implies dependency, as is so often charged, or the unique dominance of any one country such as the United States. Here let us assume that global electronic communications promote, as I have been arguing, both cosmopolitan culture and diversity and less coherence within cultures. What that would suggest is increased complexity of social structure across geographic lines, with national boundaries playing a lesser role in defining the structural lines. If there are global networks of people whose interests have become well differentiated, they may react with each other in their own invisible colleges.

Language, of course, is a serious barrier to such supranational interactions. As a barrier it will not vanish in any period worth considering, though one can expect that much thought and attention will be given to the problems that it generates. If, for example, there is a network of econometric modelers among planning commissions, one would expect much use of the English language, but also much resentment of the barriers and stratification that imposes. An intellectual subject of growing concern is likely to be the conquest of linguistic barriers by automatic and language-independent means.

A greater diversity within societies and an increase in specialized rather than mass communication might conceivably have the effect of reducing the marked propensity of modern societies to fads. The extraordinary speed and scope with which a new song, game, or hairdo sweeps across a society in the twentieth century may have something to do with the dominance of television and a few great magazines. Fads are not new; they existed before the printing press. But the rapidity with which a new heresy or faith spread in that era was fast only by the standards of that day. We can speculate about how sensitive faddism may be to any change in the structure of communications.

One thing that will not change in a world of vastly increased interpersonal communications networks is that the day has only 24 hours. The physical limitations of geography were once a tremendously important filter. A king or president could never see all those persons who wished to talk to him. Shakespeare has Julius Caesar rejecting the soothsayer who would have his ear as he walked toward the Senate House. But even if not everyone got through, a substantial proportion of those few who traveled far to make contact could get through. Even 40 years ago a long-distance call was almost a command performance. There was a man in the 1930s whose hobby was phoning the top heads of state around the world to plead with them for peace; he got through to most of them. Today, the busy leaders in the United States or other countries could not conceivably accept all their long-distance calls. There must be a filter to replace the natural geographic filter. What the etiquette and mores will be that will reestablish control of time and contacts by those who in principle are reachable by anyone we cannot forecast, but that the problem has to be solved is clear.

The new technologies of communication are in many ways very democratic. Millions of people can all take advantage of very inexpensive facilities to try to make themselves heard. The man who now laboriously writes a letter to his Congressman could on a computer network send an instantaneous personalized message to every member of Congress. It would, of course, not be listened to; it would perhaps have less effect than his present letter. Somehow, by a combination of devices and customs, a protective shell will be created as fast as the flow of communication threatens to become overwhelming. The protective customs may not be democratic; they may indeed be a reaction to that. So once again, there is no easy way to say how different societies will solve the problem at different times; what we can do is point to the problem on the horizon.

Of all the possible cultural effects of evolving communications technologies, the one on which I should like to speculate in closing this chapter is the increase in diversity in society and the breakdown of cohesion that may go with that. We can expect that there will be a great growth of specialized intellectual subcultures. There will be operas and opera news available for opera lovers, microbiology information bases and exchanges available for microbiologists. All of these will draw some portion of people's time and attention away from the common concerns of the nation's sports, politics, heroes, and news. If that happens, the complaints we would hear from social critics will be just the opposite from today's. Today the usual social criticisms of our mass-media-dominated society is that it suffers from a dull conformity. Dissent and criticism have little voice.

The technology of dynamic information handling is likely to produce opposite complaints. We are likely to hear complaints that the vast proliferation of specialized information serves only special interests, not the community. That they fractionate society, providing none of the common themes of interest and attention that make a society cohere. The critics will mourn the weakening of the national popular culture that was shared by all within the community. We will be told that we are being deluged by undigested information on a vast unedited electronic blackboard and that what a democratic society needs is shared organizing principles and consensus in concerns.

Like the present criticism of mass society, these criticisms will be only partly true, but partly true they may be. A society in which it becomes easy for every small group to indulge its tastes will have more difficulty mobilizing unity. A society where mass publishing has to compete with specialized information resources will have more trouble establishing coherence of intellectual debate.

We can speculate about possible consequences. Will there be a continuing decline of interest in national elections and a rise of interest in small group politics? Will there be a decline of popular interest in national events? Will there be a growth of esoteric sects? These are the kinds of questions that need to be asked in this context of the technological trends that we can anticipate. We may suspect that it will promote individualism and will make it harder, not easier, to govern and organize a coherent society.

Notes

1. From Mass Media Revolution to Electronic Revolution

1. Harold A. Innis, *The Bias of Communication* (Toronto: University of Toronto Press, 1951), pp. 18–19. The following brief account of the history of printing repeats information that appears in Ithiel de Sola Pool, *Technologies of Freedom* (Cambridge: Harvard University Press, 1983).
   International Communications and Information. Report to the Subcommittee on
2. *Encyclopaedia Britannica*, 15th edition, "Printing."
3. From the label in the Gutenberg Museum in Mainz.
4. Denys Hay, Introduction to John Carter and Perry Muir, *Printing and the Mind of Man* (New York: Holt, Rinehart and Winston, 1967), p. 742.
5. Elizabeth Eisenstein, "Some Conjectures about the Impact of Printing on Western Society and Thought: A Preliminary Report," *Journal of Modern History*, 40, no. 1 (1968): 1–56.
6. Ibid., p. 22.
7. That figure is derived from the Venice statistic above. What is not clear is whether helpers are included in that figure. If not, the figure might be as few as two volumes a week instead of one a day, but the conclusion is unaffected.
8. Eisenstein, "Some Conjectures," pp. 1–56.
9. *Encyclopaedia Britannica*, 15th edition, "Printing."
10. Harold A. Innis, *Empire and Communications* (Toronto: University of Toronto Press, 1972).
11. "Selected Financial and Operating Data," *Annual Report of the Postmaster General, 1939* (Washington, D.C.: U.S. Printing Office, 1939).
12. We might note that the fullest development of the arts of oral communication came only a couple of thousand years *after* writing had already been invented. The art of rhetoric and drama in Greece or that of prophecy in Palestine, even though oral, depended upon the continuity of the written tradition to achieve their full greatness.
13. Jean Gottmann, *Megalopolis: The Urbanized Northeastern Seaboard of United States* (New York: Twentieth Century Fund, 1961), p. 576.
14. Mark V. Porat and Michael R. Rubin, *The Information Economy*, 9 vols. (Washington, D.C.: Government Printing Office, 1977).
15. Cf. Max Horkeimer and Theodor W. Adorno, *Dialectic of Enlightenment* (1947; rpt. New York: Seabury Press, 1972).
16. *New York Times v. United States*, 403 U.S. 713, 91 S. Ct. 2140, 29 L. Ed. 2d 822 (1971).

17. In John Kimberly Mumford, "This Land of Opportunity," *Harpers Weekly*, 52 (August 1, 1908): 23.

2. The New Communications Technologies

1. Teletypes generally use a 5 or 6 bit representation of each letter, which allows 32 or 64 different characters. Upper and lower case full text systems may use up to 8 bits per character, or 256 different symbols. In addition, a so-called check bit for error reduction is often used. So it takes 5 to 9 bits to represent a character.
2. *United States v. Zenith Radio Corporation et al.*, 12f. 2d 614 (N.D. Ill.), April 16, 1926.
3. But cf. Ithiel de Sola Pool et al., eds., *Handbook of Communications* (Chicago: Rand McNally, 1973), pp. 462–511, for the analysis of a reverse situation. The cost of wired radio in China is less than that for battery-operated transistor sets. Battery costs make the difference. Similarly in the West in the early decades of the century, radio broadcasting was uneconomical as long as radios had to have expensive heavy batteries. Radios became practical for a broadcast system only when the improvement of tubes made it possible to work a radio off regular current.

3. Crumbling Walls of Distance

1. See in Marion May Dilts, *The Telephone in a Changing World* (New York: Longmans, Given and Co., 1941), p. 20.
2. See Ithiel de Sola Pool et al., "Foresight and Hindsight: The Case of the Telephone," *The Social Impact of the Telephone* (Cambridge: MIT Press, 1976), p. 156.
3. See Herbert N. Casson, "The Telephone as It Is Today," *World's Work*, 19 (1910): 12775.
4. See Roger Burlingame, *Engines of Democracy* (New York: C. Scribner, 1940), pp. 118f.
5. *Electrical Review*, 18, no. 7 (April 11, 1901): 98.
6. Strictly speaking, satellite communications costs are not totally insensitive to distance. By focusing the beam, one can economize on power, thus lowering costs. If, instead of broadcasting a global beam, one sends a beam that covers, say, a diameter of 1,300 miles, then the same transmission power from the satellite will produce a much stronger signal on the ground and therefore permit use of a smaller antenna. Also, if one extends the distance of a transmission to two hops, obviously the costs go up. Nonetheless, it is true that within a given beam "footprint" it makes no difference in cost how close or how far apart the points are between which the message travels.
7. Our subject here is cost. For voice transmission, telephone companies prefer to use terrestrial circuits to avoid the ¼ second delay that results from the 45,000 miles a satellite message travels.

4. Limits to Growth

1. The Sloan Commission was established by the Alfred P. Sloan Foundation in 1970 to explore the possibilities of Wired Broadband Communications (Cable Television) as a public interest service. Their conclusions and recommendations for achieving maximum potential were published in *On the Cable: The Television of Abundance* (New York: McGraw-Hill Book Company, 1971).
2. W. P. Banning, *Commercial Broadcast Pioneer: The WEAF Experiment, 1922–1926* (Cambridge: Harvard University Press, 1946).
3. Indeed, efficient use of the spectrum would only increase the number of possible competitors, something the present licensees have no interest in. They have no interest, for example, in the VHF drop-in proposal for 200 more stations.
4. Charles Jackson, *Technology for Spectrum Markets* (Cambridge: MIT Press, 1976).
5. *NBC v. United States*, 319 U.S. 190, 87 L. ed. 1344 (1943).
6. One thousand conversations at 120 words a minute, compared with 100,000 words of text a second for 60 seconds.

5. Talking and Thinking among People and Machines

1. See Raymond A. Bauer in *Handbook of Communication*, ed. Ithiel de Sola Pool et al. (Chicago: Rand McNally, 1973). See also Ithiel de Sola Pool, Suzanne Keller, and Raymond A. Bauer, "The Influence of Foreign Travel on Political Attitudes of American Businessmen," *Public Opinion Quarterly*, 20 (1956): 161–175; Ithiel de Sola Pool and Irwin Schulman, "Newsmen's Fantasies, Audiences, Newswriters," *Public Opinion Quarterly*, 23 (1959): 145–155; and Claire Zimmerman and Raymond A. Bauer, "The Effect of an Audience on What Is Remembered," *Public Opinion Quarterly*, 20 (1956): 238–248.
2. For an earlier discussion see Joseph Weitzenbaum, *Computer Power and Human Reason: From Judgement to Calculation* (San Francisco: W. H. Freeman and Company, 1976).
3. The French word reflects a somewhat different conception. Lacking a good French name for its devices, IBM turned to Professor J. Perret of the Sorbonne, who suggested the name "ordinateur." That was a theological word which had fallen into desuetude for six centuries. "God was the great 'ordinateur' of the world; that is to say the one who made it orderly according to a plan." Jean-François Blondeau et al., *Communications et société* (Paris: Institute de Recherche et d'Information Socio-Economique, Universite de Paris-Dauphine, n.d.), p. 12, citing C. Alavoine through F. Truchet, *Information et l'ordinateur*.
4. A turning point was the work of Ivan Sutherland in mid-1960s at MIT on information retrieval from a remote idea bank via the telephone network.
5. See Anthony Oettinger, Principal Investigator, *Mathematical Linguistics and Automatic Translation: Report #NSF-17 to the NSF* (Cambridge, Mass.: Computation Lab, Harvard University, 1966).

6. Claude Shannon and Warren Weaver, *Mathematical Theory of Communication* (Urbana: University of Illinois Press, 1949).
7. The Advanced Research Projects Agency Network is a National Packet Data Network developed by the Department of Defense. It is available for private use and is primarily used for file transfer and electronic mail.

6. Communities without Boundaries

1. Ralph Linton, *The Study of Man* (New York: Appleton Century Crofts, 1936), p. 325.
2. This phrase is used by Suzanne Keller in her essay "The Telephone in New (and Old) Communities" in Ithiel de Sola Pool, *The Social Impact of the Telephone* (Cambridge, Mass.: MIT Press), pp. 281–299.
3. Harold Nicholson, however, points out that far from expanding the initiative of the ambassador in those days, "most ambassadors during the period of slow communication were so terrified of exceeding their instructions . . . that they adopted a purely passive attitude." *The Evolution of Diplomatic Method* (London: Constable, 1954), p. 898.
4. Arthur C. Corte and Colin J. Warren, *Policy Issues and Data Communications for NASA Earth Observation Missions until 1985*, Center for Policy Alternatives and Center for International Studies, MIT, November 1975.
5. Herbert Casson, "The Social Impact of the Telephone," *The Independent*, 71 (Oct. 16, 1911): 901.
6. Ronald M. Westrum, "The Historical Impact of Communication Technology on Organization," Working Paper no. 56, Institute for the Study of Social Change, Purdue University, mimeo, n.d. (1976), quoting Fritz Redlich, 1952.
7. Many studies have been done by governments to examine the efficiency of moving offices out of the capital and keeping in touch with them by telecommunications. These studies have been reviewed by Alex Reid in Pool, *Social Impact of the Telephone*, pp. 386–415.
8. Marshall McLuhan, *Understanding Media: The Extension of Man* (New York: New American Library, 1964), p. 238.
9. "Congressalia," *International Association,* no. 4 (1974): 243.
10. To be more precise, the general news business grew to be much larger than the commercial services until the 1960s, when the demand for international financial information burgeoned and reversed that relation.
11. At that point business leases of such private high-speed lines were 152. Eighteen months earlier, in 1972, the figures were 78 low-speed teletype circuits and 4 high-speed lines for the press and 73 high-speed lines for business. The takeoff in use of high-speed circuits began in the multinationals a few years earlier than the press.
12. William H. Read, "U.S. Private Media Abroad" in *The Role and Control of International Communications and Information. Report to the Subcommittee on International Operations of the Committee on Foreign Relations, U.S. Senate, 95th Congress, 1st Session* (Washington, D.C.: Government Printing Office, 1977), p. 9.
13. Efforts by American magazines at foreign sales began just before World

War II. *Reader's Digest* began British sales in 1938 and founded its Spanish language edition in 1940. *Time* began overseas distribution in 1941. See Read, ibid., p. 157.

14. As with all such inventions, we could discuss a half-century history of anticipations and parallel inventions.

15. Thomas H. Guback, "Film as an International Business," *Journal of Communication*, 24 (Winter 1974): 92.

16. See Ithiel de Sola Pool, "Totalitarian Communications," in *Handbook of Communications* (Chicago: Rand McNally, 1973).

17. See W. Phillips Davison, *Mass Communication and Conflict Resolution* (New York: Praeger, 1974), for a discussion of the importance of international communications in providing a sense of world support to political movements under stress.

18. Don D. Smith, "America's Short Wave Audience: Twenty-five Years Later," *Public Opinion Quarterly*, 33 (1969–70): 537–545.

19. Tapio Varis, "International Inventory of Television Programme Structure and the Flow of TV Programmes between Nations," no. 20 (Tampere, Finland: University of Tampere, 1973). Reprinted in *Journal of Communication*, 34, no. 1 (1984): 143–152.

20. One should nevertheless not underestimate the importance of spillover across the frontiers. Since most of the Canadian population is strung along the long U.S. frontier, about one fifth of Canadians can receive U.S. broadcasts directly. Similarly, many Estonians can tune into Finnish TV; Czechoslovakians can receive Vienna and East Germans West German stations.

21. The argument is sometimes made that poor countries do not have that choice, and that American or other foreign programming is in some way forced upon them as a kind of cultural imperialism. (See Herbert Schiller, *Communication and Cultural Domination*, White Plains, N.Y.: International Arts and Science Press, 1976.) It is true that commercial producers of entertainment serials try to pick up added revenue by renting returns at whatever price the international market will bear, and that means at very low rates to poor countries. As a result, developing countries often rent U.S. serials of a sort that their intellectuals and politicians attack as irrelevant to their needs and harmful to their culture. They rent them partly because in underdeveloped countries there are also underdeveloped bureaucracies; often the left hand does not know what the right hand is doing. An official in the broadcasting organization may contract for a program that a salesman tells him other countries have shown with popular success; at the same time his delegate to the UN may be making an impassioned speech about the evils of sex, violence, and commercialism on exported TV. The reason for renting American commercial programs is not a lack of choice. If one wants more elevated materials, there are many low-cost documentaries from good causes and international organizations. If one wants material more appropriate to poor countries, there are plenty of tapes on health and agriculture, including many from the United States. If one wants material that is not from a capitalist point of view, there are many Soviet and East European materi-

als. If the answer is given that such programs are relatively dull, and that the commercial material is more exciting, one can usually agree. That, however, raises issues far more fundamental than any charge of cultural imperialism. It leaves the eternal question that is central to education and culture in all societies: how to make good things enjoyable.

22. Raymond A. Bauer, Ithiel de Sola Pool, and Lewis A. Dexter, *American Business and Public Policy* (Chicago: Atherton-Aldine, 1963).

23. In Chapter 11 we will look at the communication/transportation trade-off. The evidence seems to suggest that the price elasticity of international telephone calls is somewhere between $-.6$ and -1.0.

24. Irving Janis, *Groupthink: Psychological Studies of Policy Decisions and Fiascoes* (Boston, Mass.: Houghton Mifflin, 1982).

25. The proceedings of regulatory agencies are sufficiently similar to computer conferencing to suggest a possible application. In such proceedings at present, parties file comments, reply-comments, replies to replies, and so forth. Each party generally mails a copy of his submission to all of the other parties. The general public, however, may have to visit the regulatory agency headquarters or large public libraries to examine these master filings. As the contents of such submissions are almost exclusively text, putting the record of such proceedings on a computer would present few problems. Interested members of the public could then have easy access to the record. There would be savings in cost and the potential to expedite such proceedings, which often take years.

26. Murray Turoff, *Computerized Conferencing and Real Time Delphis, Unique Communication Forms*, Proc. 1st International Computer Communications Conference, p. 135. In these cost comparisons, computer conferencing costs of $7 per hour and two assumptions concerning the value of participant time were used. The inclusion of travel costs in the comparison would make the case for computer conferencing even more convincing.

27. H. G. Wells, "The Brain Organization of the Modern World," lecture, October and November 1937.

28. Edward Chamberlin in *The Theory of Monopolistic Competition* (Cambridge: Harvard University Press, 1939), p. 9, defines monopolistic competition as competition between sellers of differentiated products. "With differentiation appears monopoly, and as it proceeds further the element of monopoly becomes greater. Where there is any degree of differentiation whatever, each seller has an absolute monopoly of his own product, but is subject to the competition of more or less imperfect substitutes."

29. For example, every time a stock price changes in the files of a business information service it needs to be changed not only in the daily stock tables but in the information files on that company, in the daily average, and any other place where it appears.

7. Regulating International Communication

1. Ithiel de Sola Pool, *Technologies of Freedom* (Cambridge, Mass.: Harvard University Press, 1983).

2. Eli M. Noam, *Telecommunications in Europe* (Cambridge: Harvard University Press, forthcoming), and Eli M. Noam and Seisuke Komatsuzaki, eds., *Telecommunications in the Pacific Basin* (forthcoming).

3. The United States used to apply the same distinction for international traffic. Until 1976 the FCC prohibited use of AT&T international lines for transmission of data, thus making uneconomical the legal use of facsimile, acoustic couplers, or computer polling systems by anyone except big institutions with leased lines.

4. "Treaty on Principles Governing the Activities of States in the Exploration and Use of Outer Space, Including the Moon and Other Heavenly Bodies," adopted by the UN General Assembly, 27, Jan. 6, 1967.

5. The United States never ratified it on the grounds that it also contains some clauses against international propaganda which would imply control of free speech. It would be more accurate, however, to say that excuse was used to avoid coming to the Senate with a treaty that would be blocked, though more by nationalist conservatives than by libertarians.

6. At the Nairobi UNESCO general conference in 1976 a Soviet-backed resolution with those words was shelved for two years because of Western opposition.

7. The Human Rights Convention contains a clause prohibiting racist or war propaganda. That is the clause that the United States said violated the First Amendment.

8. This extreme view was not accepted as a norm in the creation of the news agency pool for nonaligned countries at the New Delhi conference of July 1976. Many countries, of course, have such agencies. India, for example, under Indira Gandhi compelled the merger of all press agencies into the single national agency, Samachar. The conference documents avoided telling member countries how to organize their media. Nonetheless, the valid fear of liberals and the world press was that a number of countries would follow the encouragement to form national news agencies to the Indian conclusion of excluding any independent ones.

9. One practice of monopsonistic carriers is to purchase their equipment from their own subsidiary. Where governments do not allow such procurement, or where the phone system is too short of capital or other resources to do their own manufacturing, the phone system generally buys from a small set of suppliers with whom it has developed intimate understandings and relations. Other firms set up manufacturing subsidiaries in those countries they hope to sell to. The result is a quite limited amount of foreign trade in telecommunications products for most countries relative to the use of the telecommunications systems.

10. Raymond Vernon, *Sovereignty at Bay* (New York: Basic Books, 1971); Raymond Vernon, *The Economic Environment of International Business* (Englewood Cliffs, N.J.: Prentice Hall, 1972); Robert Stobaugh, *Nine Investments Abroad and Their Impact at Home* (Cambridge, Mass.: Harvard University Press, 1976).

11. The process in the electronics industry is described in a report: "Domestic

Employment Effects of Foreign Direct Investment in the Electronics Industry," Center for International Studies, MIT, January 1975.

12. David R. Simon, book review summarizing the argument of Herbert I. Schiller, *Communication and Cultural Domination* (White Plains, N.Y.: International Arts and Science Press, 1976). *Journal of Communication*, 27, no. 2 (Spring 1977): 22.

13. Elihu Katz et al., eds., *Communications Policy for National Development* (London: Routledge and Kegan Paul, 1977). The terms "epochalism" and "essentialism" are from Clifford Geertz, *The Interpretation of Cultures* (New York: Basic Books, 1973). In Katz's words: "Geertz describes the dilemma as finding a proper balance between the cosmopolitan, future-oriented 'spirit of the age'—what he calls 'epochalism'—and the common experiences that inhere in tradition—or 'essentialism'. The sources of the latter, says Geertz, are parents, traditional authority figures, custom and legend, and of the former they are secular intellectuals, the oncoming generation, current events, and the mass media. For Geertz, the mass media seem inexorably opposed to 'essentialism'."

14. Daniel Lerner, *The Passing of Traditional Society* (Glencoe, Ill.: Free Press, 1958).

15. Herbert Schiller, *Mass Communication and American Empire* (Glencoe, Ill.: Free Press, 1958). See also his *Mind Managers* (Boston, Mass.: Beacon Press, 1973) and *Communication and Cultural Domination* (White Plains, N.Y.: International Arts and Sciences Press, 1976).

16. See Georgi Arbatov, *The War of Ideas in Contemporary International Relations* (Moscow: Progress Publishers, 1973), for a Soviet statement.

17. Katz, *Communications Policy;* see also Katz and George Wedell, *Broadcasting in the Third World: Promise and Performance* (Cambridge, Mass.: Harvard University Press, 1977).

18. *Intermedia*, no. 3, 1973.

19. Ibid.

20. Ibid.

21. Karl Marx and Friedrich Engles, *The Communist Manifesto* (New York: Monthly Review Press, 1964), p. 35.

22. Mr. Christopher Nacimento draws that conclusion: The alternative for many a poor developing nation [to the presence and power of foreign owned multinational mass media empires] is going to have to be, at least for a time, government ownership of the mass media . . . Insofar as governments can claim to speak for the people, broadcasting in a developing country cannot fall outside government policy." *Intermedia*, no. 3, 1973.

23. William Read, "U.S. Private Media Abroad" in *The Role and Control of International Communications and Information. Report to the Subcommittee on International Operations of the Committee on Foreign Relations, U.S. Senate, 95th Congress, 1st Session* (Washington, D.C.: Government Printing Office, 1977), p. 9.

24. *The Study of Man* (New York: Appleton Century Crofts, 1936), p. 325.

25. Another reason is that much research in the United States has been

federally funded, and a condition is that licenses be granted for government use.

26. Karl W. Deutsch, "Shifts in the Balance of Communications Flows," *Public Opinion Quarterly*, 20 (1956): 143–160.
27. *The Media Are American* (New York: Columbia University Press, 1977), p. 3.
28. For a review of the economics of international media flows see Joel Millonzi and Eli M. Noam, eds., *International Trade in Film and Television* (Norwood, NJ: Ablex, forthcoming). Also, Eli M. Noam, *Television in Europe* (Cambridge: Harvard University Press, forthcoming).
29. Derek J. de Solla Price, *Little Science, Big Science* (New York: Columbia University Press, 1963).

8. Broadcasting from Satellites to Home Receivers

1. Quoted in Albert Parry, *Russia's Rockets and Missiles* (Garden City, N.Y.: Doubleday, 1960), pp. 208–209.
2. James Dukowitz, "The Grand Negotiation: Politics, Satellites, and Decision-Making in the ITU," unpublished diss., MIT, 1973.
3. Ibid., p. 38.
4. Since countries sufficiently remote from each other can use the same frequency without interference, the product of the number of countries multiplied by five is higher than forty.
5. Jamming of a highly directional signal like that from a satellite is not easy except by another highly directional signal. Ground wave jamming would not be very effective.

9. Communications for the Less Developed Countries

1. See "Passing of the Telephone Girl," *Harpers Weekly*, 53 (Sept. 18, 1909): 37, describing Germany.
2. It had in that era just converted at great cost to French central power phones. But once introduced in the U.S.A., automatic switching became almost universal, and the Bell System jumped ahead once more by making long-distance dialing automatic throughout the country.
3. I. W. Aitkin, "Automatic Telephone Exchange Systems," *Science*, American Supplement, vol. 73, Jan. 13, 1912, pp. 20–22.
4. Obviously there are other reasons, too, such as low labor costs attracting capital to a country.
5. Fredrick Frey, "Communications and Development," in *Handbook of Communication*, ed. Ithiel de Sola Pool et al. (Chicago: Rand McNally, 1973).
6. Daniel Lerner, *The Passing of Traditional Society* (Glencoe, Ill.: The Free Press, 1958).
7. Hayward Alker, "Causal Inference and Political Analysis," in Joseph Bernd, ed., *Mathematical Applications in Political Science* (Dallas: Southern Methodist University Press, 1966). Phillips Cutright, "National Political

Development: Measurement and Analysis," *Am. Sociological Review*, 28 (April 1963): 253–264. Donald McCrone and Charles F. Crudde, "Toward a Communications Theory of Democratic Political Development: A Causal Model," *American Political Science Review*, 61 (March 1967): 72–79. Gilbert Winham, "Political Development and Lerner's Theory: Further Tests of a Causal Model," *American Political Science Review*, 64 (September 1970): 810–811.

8. Ithiel de Sola Pool, "The Mass Media and Politics in the Modernization Process," in *Communications and Political Development*, ed. Lucian Pye (Princeton: Princeton University Press, 1963).

9. A repairman may often use the telephone that way in the United States where it is cheap and works well. Developing countries cannot hope to have that kind of telephone service at prices they can afford within the next couple of decades, but low-bandwidth text and data communication can make the same kind of interaction available at minimal costs on both a national and international basis.

10. See Asghar Fathi, "The Minbar as Medium of Public Communication in Islam," *Communications and Development Review*, Iran Communications and Development Institute, Teheran, 1, no. 1 (Spring 1977): 11–12; Benker Roy, "Communication Challenge in Rural Areas," *Indian Express*, January 19, 1977.

11. Lerner, *Passing of Traditional Society.*

12. Among the more important historical treatments are: V. I. Lenin, *What Is To Be Done*, trans. J. Fineberg and G. Hanna, ed. V. J. Jerome (New York: Internation Publ., 1968); Roberto Michels, *Political Parties*, trans. Eden and Cedar Paul (Glencoe, Ill.: Free Press, 1958); Max Weber, *The Theory of Social and Economic Organization*, trans. A. M. Henderson and Talcott Parsons (New York: Oxford University Press, 1947); M. Ostrogorskii, *Democracy and the Rise of Political Parties*, trans. Fredrick Clarke, ed. J. Bryce (Chicago: Quadruple Books, 1964).

13. See Vance Packard, *The Hidden Persuaders* (New York: D. McKay, 1957).

14. Everett Rogers and F. Floyd Shoemaker, *Communication of Innovations: A Cross Cultural Approach* (New York: Free Press, 1971).

15. Ibid.

16. Elihu Katz and Paul Lazarsfeld, *Personal Influence* (New York: Free Press, 1964).

17. Lerner, *Passing of Traditional Society*; David Clarence McClelland, *The Achieving Society* (London: Collier-Macmillan, 1961); and Alex Inkeles, *Exploring Individual Modernity* (New York: Columbia University Press, 1983).

18. I include under the heading of religious movements such secular religions as Maoism. The distinctive characteristics of such movements, besides demanding that the members change their lives to conform to a code of new rituals and customs, include membership in a tight-knit group which reinforces the preachings on an intimate and day-by-day basis, and also mystical beliefs and sanctions. The great world religious movements like Christianity, the Reformation, and Islam are obvious

examples, but besides these, throughout history there have been smaller local evangelical movements that uplifted the life of the true believers and produced significant reform in their society.

19. M. S. Gore, *The SITE Experience Reports and Papers on Mass Communications*, no. 91, UNESCO, undated.

20. See Ithiel de Sola Pool, "The Public and the Polity," in Pool, ed., *Contemporary Political Science: Toward Empirical Theory* (New York: McGraw-Hill, 1967), pp. 22–52.

21. Charles Murray, *A Behavioral Study of Rural Modernization: Social and Economic Change in Thai Villages* (New York: Praeger, 1977).

22. Such papers have been particularly important in Africa because there is much less heritage of a regular press there than there is in Asia or Latin America.

23. Alan O. Liu, *Communications and National Integration in Communist China* (Berkeley: University of California Press, 1971).

24. Henry Cassirer, *Television Teaching Today* (Paris: UNESCO, 1960).

25. The comparison is somewhat more complicated. The mean time to failure for electronic equipment is likely to be longer than that for mechanical, but when failures do occur with electronic equipment they are apt to be total, while with mechanical equipment they are partial.

26. See Lerner, *Passing of Traditional Society*; Alex Inkeles, *Becoming Modern* (Cambridge, Mass.: Harvard University Press, 1974).

27. Max Weber, *The Protestant Ethic and the Spirit of Capitalism*, trans. Talcott Parsons (New York: Scribner's, 1930).

28. Ithiel de Sola Pool, *Democracy in a World of Tensions*, a UNESCO symposium, ed. Richard McKeon and Stein Rokkan (Chicago: University of Chicago Press, 1951), pp. 328–352.

10. Advanced Communications and World Leadership

1. See Otto Reigel, *Mobilizing for Chaos* (New Haven: Yale University Press, 1934).

2. Erik Barnouw, "A History of Broadcasting in the United States," in *A Tower in Babel*, vol. 1 (New York: Oxford University Press, 1966), pp. 53–60.

3. *The Times*, June 22, 1977, p. 22.

4. Daniel C. Roper, *The United States Post Office* (New York: Funk and Wagnalls, 1917), p. 9.

5. One example is the formation of RCA under pressure from the U.S. government, which did not wish transatlantic radio telegraphy to be monopolized by the British Marconi Company.

6. Barton Whaley, *Codeword BARBAROSSA* (Cambridge, Mass.: MIT Press, 1973).

7. Eli M. Noam, "The Public Telecommunication Network: A Concept in Transition," *Journal of Communications*, 37, no. 1 (Winter 1987): 30–48.

11. *The Ecological Impact of Telecommunications*

1. "Telephone and the Doctor," *Literary Digest*, 44 (May 18, 1912): 1037. Before the telephone, it was practical for doctors to be in a single block; if the physician one sought was away, another was nearby. After the introduction of telephones one could locate the physician one wanted.

2. J. Alan Moyer, "Urban Growth and the Development of the Telephone: Some Relationships at the Turn of the Century," *The Social Impact of the Telephone*, Ithiel de Sola Pool, ed. (Cambridge: MIT Press, 1978). For London see P. M. Townroe, *Social and Political Consequences of the Motor Car* (Newton Abbot: David and Charles, 1974), pp. 52f: "One can illustrate this by looking at the development of the built-up area of London from the seventeenth century. One must realize that at that time there was no cheap public transport . . . In the eighteenth century a horse cost about three times as much as a motor car costs today in terms of working-class wages. A stage coach from Paddington into the city cost 2s 0d— single, which was 10 percent of a typical working-class wage. A taxi to do the same trip today would cost 2 percent of a typical working-class income, and a bus would be much cheaper. The only mass transport was on foot, and the cities were therefore very dense because a person had to live within a mile or so of his place of employment. In London, because there was some transport along the river, there was a linear development about a mile deep along the banks, particularly on the Northern side.

   As we go on 130 years to 1800, we find that the built-up area has not expanded much . . .

   By 1950, only 150 years later, the built-up area had grown by a linear factor of about 10. This . . . was the result of a whole series of transport improvements during the nineteenth century and the first half of the twentieth century. These started in 1829 with the horse buses . . . These were followed by the urban railways from 1836 onwards. By 1880 there existed an area of central development, which was being served by horse buses and horse trams, plus development along the railway lines."

3. Jean Gottmann, "Megalopolis and Antipolis: The Telephone and the Structure of the City," in Pool, *Social Impact of the Telephone*.

4. H. G. Wells, *Anticipations* (New York: Harper Bros., 1902), p. 51.

5. Ibid., pp. 52f.

6. Ibid., p. 58.

7. Ibid., p. 66.

8. "Action at a Distance," *Scientific American*, 77 (1914): 39.

9. While in general the phone company supported the zoning movement, on one point it was at odds with the urban planners. Zoning was often used to set height limits, restricting the construction of tall buildings. Those, however, were heavy telephone users. Hence, telephone companies opposed height limitations.

10. See President's Research Committee, *Recent Social Trends* (New York: McGraw-Hill, 1933), pp. 197ff.

11. Arthur Pound, *The Telephone Idea: Fifty Years After* (New York: Greenberg, 1926).
12. Theodore K. Noss, *Resistance to Social Innovations* (Chicago: University of Chicago, Dept. of Sociology, 1944).
13. Vol. 52, Sept. 20, 1879, pp. 1187–1188.
14. "The Telephone in Modern Business," *Telephony*, 2, no. 5, pp. 65f.
15. Herbert N. Casson, "The Telephone as It Is Today," *World's Work,* 19 (April 1910): 12777.
16. See John Short, Ederyn Williams, and Bruce Christie, *The Social Psychology of Telecommunications* (London: John Wiley, 1976), pp. 10–11, 166–167. They report an interesting example of synergistic effect in the reverse direction. "The opening of the Severn Bridge, a road bridge linking Southwest England with South Wales, was soon followed by the jamming of the telephone trunk routes between those areas" (p. 11).
17. See "Sociological Effects of the Telephone," *Scientific American*, 94 (1906): 500.
18. Bertil Thorngren, "Silent Actors: Communication Networks for Development," in Pool, *Social Impact of the Telephone.*
19. Alphonse Chapinis et al., "Studies in Interactive Communication: The Effects of Four Communication Modes on the Behavior of Teams during Cooperative Problem Solving," *Human Factors*, 14 (1972): 487–509.
20. Alex Reid, "Comparing Telephone with Face-to-Face Contact," in Pool, *Social Impact of the Telephone.* See also Short, *Social Psychology of Telecommunications.*
21. Thorngren, in Pool, *Social Impact of the Telephone.*
22. *Long Range Intelligence Bulletin 7*, Post Office Telecommunications, January 1976.
23. Ibid., p. 35.
24. Pauline Wingate, "Newsprint: From Rags to Riches—And Back Again," in Anthony Smith, ed., *Newspapers and Democracy* (Cambridge: MIT Press, 1980), p. 67; David C. Smith, "Wood Pulp & Newspapers, 1867–1900," *Business History Review*, 3 (Autumn 1964): 328–345.
25. "To Prolong Life of Telephone Poles," *Telephony*, 10, no. 1 (July 1905).
26. GNP is based upon market price. As a resource becomes more scarce, its contribution to GNP goes up; as it becomes plentiful, its price and contribution to GNP goes down. The information sector has been growing so rapidly that even despite falling prices its contribution to GNP has gone up. In the future, real information and communication may grow; whether its contribution to GNP will grow is another question.

12. Technology and Culture

1. Michael Robinson, *Over the Wire and on TV* (New York: Russell Sage Foundation, 1983).
2. See S. C. Gilfillan, *Invention and the Patent System*. Printed for the use of

the Joint Economic Committee (Washington, D.C.: U.S. Government Printing Office, 1969), pp. 110–111.

3. Annan Commission, *How to Organize TV4: Report of the Committee on the Future of Broadcasting*, HMSO CMND 6753 (London, March 1977).

4. This proposition is supported in the literature in Hilde T. Himmelweit, *Television and the Child* (London: Oxford University Press, 1958), and Bernard Berelson and Gary Steiner, *The People Look at Television* (New York: Knopf, 1963).

5. For an economic analysis of these issues, see Eli M. Noam, "A Public and Private-Choice Model of Broadcasting," *Public Choice*, 55 (1987): 163–187.

6. U.S. Cabinet Committee of Cable Communications, *Cable: Report to the President* (Washington, D.C.: USCPO, 1974).

7. One possible difference is geographic. Any individual can receive by mail a paper or magazine from another city or from abroad. But paper is bulky and expensive to ship, which gives the local press a competitive edge. The importing of electronic material for individuals would require mailing disks or cassettes. However, for relatively small groups it would be economical to import signals electronically from anywhere.

8. See Harold A. Innis, *The Bias of Communication* (Toronto: University of Toronto Press, 1951), pp. 40ff, 105ff.

9. William N. McPhee, *Formal Theories of Mass Behavior* (New York: Free Press, 1963).

10. Patrick C. Suppes and Mona Morningstar, *Computer Assisted Instruction at Stanford, 1966–68* (New York: Academic Press, 1972); Seymour Papert, *Mindstorms: Children, Computers and Powerful Ideas* (New York: Basic Books, 1980).

11. For a reposte to such assertions see Susan Holmes and Tim Rix, "Beyond the Book," in Asa Briggs, ed., *Essay in the History of Publishing* (London: Longman, 1974), pp. 319–356.

12. Florence Low, "The Reading of the Modern Schoolgirl," *Nineteenth Century*, 54 (Feb. 1906): 282, quoted in Holmes and Rix, ibid., p. 323.

13. W. K. L. Dickson, quoted in Holmes and Rix, ibid.

14. Orlo Williams, quoted in ibid., p. 326.

15. Ibid., p. 327.

16. Paul Lazarsfeld, *Radio and the Printed Page* (New York: Duell, Sloan and Pearce, 1940).

17. See O. H. Cheney, *Economic Survey of the Book Industry, 1930–31* (New York: National Association of Book Publishers, 1931).

18. The word first appears in 1767 in Blackstone's *Commentaries.* "However, the concept of copyright goes back much further than Blackstone, and its origins are now lost in the mists of time. In effect, though, the right only began to assume importance when the invention of printing made the multiplication of "copies" of a work infinitely quicker and cheaper than the painstaking products of monkish scribes, as well as appreciably more accurate than the compositions of most professional scriverners." Ian Parsons, "Copyright and Society," in Briggs, *Essays in the History of Publishing*, p. 31.

19. Ibid., pp. 33f.
20. 209 U.S. 1 (1908). See also *Goldstein v. California* 412 U.S. 546 (1973) on sound recordings. Within the context of the protection of writing, it is the form and manner of expression that is protected. Typical of the statement in case law of this principle can be found in *Becker v. Loew's Inc.*, 133 F. 2d 889 (7th Cir., 1943), certiorari denied.
21. 17 U.S.C. Sec. 101.
22. *Berlin v. E.C. Publications, Inc.* C.A.N.Y. 1964, 329 F. 2d 541, certiorari denied 85 S. Ct. 46 and others in 17 U.S.C.A., sec. 1.
23. This distinction poses some peculiar problems. For example, though a ticker service product that presented facts has been denied protection in *National Telegraph News Co. v. Western Union Telegraph Co.*, 91 F. 2d 484 (9th Cir., 1902), a newly organized telephone directory proved protectable, *Leon v. Pacific Telephone and Telegraph*, 91 F. wd. 484 (9th Cir., 1937), since the latter proved to be a new rendition of the manner in which known facts were presented.
24. Generally, distribution to the public via sale or otherwise is considered publication, but the special legal problems this concept provides are more complex than the public understands them to be.
25. See *Taylor Instrument Co. v. Fawly-Brost Co.*, 139 F. 2d 98 (7th Cir., 1943) and *Brown Instrument Co. v. Warner*, 161 F. 2d 910 (D.C. Cir., 1947) and cases cited in "Computer Software: Beyond the Limits of Existing Proprietary Protection Policy," *Brooklyn Law Review*, Summer 1973, pp. 116–146.
26. 17 U.S.C., Sec. 1 (a).

Index